Iconic Arithmetic

Volume I

Any comments, corrections, refinements or
suggestions you may have will be greatly appreciated.

I'm available via email at
william@iconicmath.com

Thanks.

Iconic Arithmetic Volume I Copyright © 2014-2019 William M. Bricken
All rights reserved.
ISBN (soft-cover edition 2): 978-1-7324851-3-6
Unary Press, an imprint of
Unary Computers, Snohomish Washington USA

Iconic Arithmetic

simple

sensual

postsymbolic

Volume I
The DESIGN of MATHEMATICS
for HUMAN UNDERSTANDING

William Bricken

in memoriam
Richard G. Shoup
George Spencer Brown

Chapters Volume I

	Preface	xxiii
1.	Context	1
2.	Ensembles	33
3.	Depth	65
4.	Dynamics	87
5.	Structure	119
6.	Perspective	133
7.	Units	161
8.	Transformation	183
9.	Accumulation	207
10.	Reflection	239
11.	Numbers	261
12.	Extension	287
13.	Dialects	305
14.	Alternatives	341
15.	Next	369

Contents Volume I

Chapters i
Table of Contents iii
List of Figures xi
Cast of Characters xv
James Algebra xxi
Preface xxiii

Chapter 1. Context — 1

   1.1 Communication — 3
         Three Volumes
   1.2 History — 5
         Belief
         Vision
   1.3 Boundary Forms — 11
         Containers
         Distinction
   1.4 Embodiment — 16
         Rigor
         Truth and Beauty
   1.5 Humane Mathematics — 20
         Simplicity
   1.6 Remarks — 25
         Endnotes

Chapter 2. Ensembles — 33

   2.1 Unit-ensembles — 34
         Units
         Accumulation
         Textual Notation

2.2 Ensemble Arithmetic   39
    Addition
    Fusion
    Polarity
    Reflection
2.3 Substitution   47
    Multiplication
    Commutativity
    Division
    Polar Multiplication
2.4 Comparison   58
2.5 Remarks   58
    Endnotes

Chapter 3. Depth   65

3.1 Order of Magnitude   67
3.2 Standardization   69
    Group
    Merge
    Canonical Form
    Reading
3.3 Parens Arithmetic   74
    Animation
    Addition
    Subtraction
    Multiplication
3.4 Remarks   81
    Endnotes

Chapter 4. Dynamics   87

4.1 Variety   88
4.2 Circle Arithmetic   89
    Fluid Boundaries
4.3 An Iconic Calculator   94
    Interface
    Modes
    Animation

4.4 Network Arithmetic   99
        Reading
        Standardization
        Addition
        Multiplication
        Arrangement
        Implementation
        Decimal Networks
4.5 Block Arithmetic   109
4.6 Viewpoint   113
4.7 Remarks   115
        Endnotes

Chapter 5. Structure   119

5.1 James Algebra   120
        Patterns and Principles
        Arithmetic
        Axiomatic Style
        Notation
5.2 Remarks   124
        Endnotes
     Concepts   126
     Axioms, Theorems   127
     Frames, Ensembles, Depth-value, Logic 128
     Maps   129
     Volume II   130
     Volume III   131

Chapter 6. Perspective   133

6.1 Diagrammatic Math   134
        Boundary Algebra
6.2 Container Types   139
        Pattern Axioms
        Partial Ordering
        Interpretation
6.3 Multidimensional Form   144
        Dialects

        Hybrid Notation
        Object/Process
6.4 Features   150
6.5 Strategy   153
6.6 Remarks   154
        Endnotes

Chapter 7. Units   161

7.1 Nothing   163
        Zero
        Much Less
        One
        Existence
7.2 Natural Numbers   170
        Accumulation
        Cardinality
        Base-free Exponents
7.3 Unification   175
7.4 Remarks   176
        Endnotes

Chapter 8. Transformation   183

8.1 Inversion   184
        Double-Boundaries
        Void-equivalent Forms
8.2 Pattern Variables   186
        Interpretation
        Alternating Contexts
        Logarithmic Foundation
8.3 Arrangement   193
        Generality
        Parallel Arrangement
8.4 Frames   198
        Frame Types
8.5 Remarks   201
        Endnotes

Chapter 9. Accumulation — 207

- 9.1 Sets and Fusions — 209
- 9.2 Counting — 211
  - Choreography
  - Domain
  - Indication
  - Replication
  - Generalized Counting
- 9.3 Multiplication — 224
- 9.4 Exponents — 227
  - Rules
  - Logarithms
- 9.5 The Accumulation Family — 231
- 9.6 Remarks — 233
  - Endnotes

Chapter 10. Reflection — 239

- 10.1 Patterns — 241
  - Dominion
  - Promotion
  - Permeability
- 10.2 Context — 247
  - Embedded Base
  - Inverse Operations
  - Reflected Forms
- 10.3 Remarks — 257
  - Endnotes

Chapter 11. Numbers — 261

- 11.1 Numerals — 261
  - Variety
  - Polynomials
  - Factors
- 11.2 Real Numbers — 271
  - The Cut
  - Computation
  - Calculation

11.3 Remarks   277
         Endnotes

Chapter 12. Extension   287

12.1 Cardinality   287
         Depth-value
12.2 Fractions   291
         Parallel Multiplication
         Reducing Fractions
         Adding Fractions
         Compound Ratios
         Adding Ratios
12.3 Bases   297
         Logarithmic Base
         Logs and Powers
         Fractional Bases
         Base-e
12.4 Fractional Reflection   301
12.5 Remarks   303
         Endnotes

Chapter 13. Dialects   305

13.1 Postsymbolic Form   306
         Dimensionality
         Varieties
13.2 Roadmap   311
         Reconstructing Brackets
         Selected Patterns
13.3 Design   314
         Bounding Box Dialect
         Bucket Dialect
         Network Dialect
         Map Dialect
         Wall Dialect
         Room Dialect
         Block Dialect
         Path Dialect

13.4	Applications	329
	Calculation	
	Derivation	
	Demonstration	
13.5	Remarks	339
	Endnotes	

Chapter 14. Alternatives — 341

14.1	Spencer Brown Numbers	342
	Arithmetic	
	Axioms	
	Issues	
	Kauffman Numbers in the Form	
14.2	Kauffman Two Boundary Form	350
	Difference	
14.3	Kauffman String Arithmetic	353
	Subtraction	
14.4	Spatial Algebra	355
	Distribution	
14.5	Models of Multiplication	361
	Grouping	
	Merging	
	Substituting	
14.6	Remarks	364
	Endnotes	

Chapter 15. Next — 369

15.1	Choice	369
	Human Nature	
15.2	A Hidden Motive	372
	Arithmetic and Logic	
	Construction	
15.3	It's Not Easy	375
	Math Education	
15.4	Summary	377
	From Here	
15.5	Remarks	380
	Endnotes	

Bibliography	383
Index to the Index	389
Index	390
People	390
Symbolic Concepts	392
arithmetic	
mathematics	
number	
representation	
Iconic Concepts	394
boundary thinking	
containment	
iconic form	
maps	
principles	
Ensembles	398
block arithmetic	
circle arithmetic	
depth-value	
ensemble arithmetic	
iconic calculator	
network arithmetic	
parens arithmetic	
pattern equations	
James Algebra	401
dialects	
James form	
pattern equations	
Other Iconic Systems	403
boundary logic	
Kauffman arithmetics	
spatial algebra	
Spencer Brown numbers	
symbols and icons	
Cover	405
Websites	406

List of Figures

Figure 2-1: *Whole numbers as ensembles* — 35
Figure 2-2: *Addition of unit-ensembles* — 41
Figure 2-3: *Identical partitions* — 42
Figure 2-4: *Integers to unit-ensembles* — 45
Figure 2-5: *Substitution* — 49
Figure 2-6: *Multiplication of unit-ensembles*, 4x3 — 51
Figure 2-7: *Options for substitution of signed units* — 54
Figure 2-8: *Axiom systems for arithmetic* — 59

Figure 3-1: *Place-value to depth-value* — 68
Figure 3-2: *Depth-value transformation patterns* — 69
Figure 3-3: *Depth-value standardization*, base-2 — 72
Figure 3-4: *Animation of parens transformations* — 75
Figure 3-5: *Parens addition*, base-10: 432+281 — 76
Figure 3-6: *Fluid circle addition*, base-10-digit: 432+281 — 76
Figure 3-7: *Parens parallel addition*, base-2: 1618+261+4401 — 77
Figure 3-8: *Parens subtraction*, base-10: 432–281 — 79
Figure 3-9: *Parens parallel addition*, base-2: 234–45+329–171–164+10 — 79

Figure 4-1: *Depth-value patterns* — 88
Figure 4-2: *Circle dialect:* 5x7 and 35÷5 — 90
Figure 4-3: *Fluid-circle dialect:* 319x548, frames 1-5 — 92
Figure 4-4: *Fluid-circle dialect:* 319x548, frames 6-10 — 93
Figure 4-5: *Iconic calculator user interface* — 95
Figure 4-6: *Iconic calculator base and space modes* — 96
Figure 4-7: *Iconic calculator animation frames* — 98
Figure 4-8: *Network numbers from* 1 to 16, base-2 — 99
Figure 4-9: *Network transformations* — 101
Figure 4-10: *Network add and multiply* — 102
Figure 4-11: *Network addition:* 5+7 — 103
Figure 4-12: *Network multiplication:* 7x5 — 104
Figure 4-13: *Network multiplication:* 5x7 — 104
Figure 4-14: *Network arrangement* — 105

Figure 4-15:	*Binomial multiplication as network arrangement*	106
Figure 4-16:	*Decimal-network addition and multiplication*	108
Figure 4-17:	*Block numbers and operations,* base-2	109
Figure 4-18:	*Block addition and subtraction,* base-2: 5+7 *and* 5–6	110
Figure 4-19:	*Block multiplication,* base-2: 7x5 *and* 5x7	111
Figure 4-20:	*Block multiplication,* base-10-digit: 319x548	112
Figure 5-1:	*James boundaries and an interpretation*	120
Figure 5-2:	*List of typographical delimiters*	124
	Concepts	126
	Axioms, Theorems	127
	Frames, Ensembles, Depth-value, Logic	128
	Maps	129
	Volume II: Equations, Angle, Nonlinear	130
	Volume III: J Patterns	131
Figure 6-1:	*James units and operations*	139
Figure 6-2:	*Pattern axioms of James algebra*	140
Figure 6-3:	*Algebraic operations to patterns of containment*	143
Figure 6-4:	*One-, two- and three-dimensional forms of multiplication*	146
Figure 6-5:	*More forms of multiplication*	147
Figure 7-1:	*Integers to James units*	171
Figure 8-1:	*Varieties of* 2x3	191
Figure 8-2:	*Types of frames*	199
Figure 9-1:	*What can be counted?*	210
Figure 9-2:	*Counting drawers*	213
Figure 9-3:	*The choreography of counting*	214
Figure 9-4:	*Accumulating and grouping tallies*	230
Figure 9-5:	*James algebra calculation*	232
Figure 9-6:	*Numeric abstraction*	233
Figure 10-1:	*Arithmetic operations and their inverses*	248
Figure 10-2:	*Reflection within arithmetic operations*	249
Figure 10-3:	*The containment structure of inverse operations*	252
Figure 10-4:	*The reflection structure of inverse operations*	252
Figure 10-5:	*Angle-bracket forms*	256

Figure 11-1: *Types of numbers* — 263
Figure 11-2: *Varieties of numeric encoding* — 265
Figure 11-3: *Factored number to depth-value* — 270
Figure 11-4: *Examples of James arithmetic calculation* — 278

Figure 12-1: *Types of cardinality* — 290
Figure 12-2: *Customized versions of arrangement as addition of fractions* — 294
Figure 12-3: *Logarithms with a range of bases* — 298

Figure 13-1: *Roadmap for generating spatial dialects* — 310
Figure 13-2: *Structural transformations for generating spatial dialects* — 311
Figure 13-3: *Pattern equations selected for display* — 315
Figure 13-4: *Bounding box dialect* — 316
Figure 13-5: *Bucket dialect* — 317
Figure 13-6: *Network dialect* — 318
Figure 13-7: *Map dialect* — 321
Figure 13-8: *Wall dialect* — 323
Figure 13-9: *Room dialect* — 324
Figure 13-10: *Block dialect* — 326
Figure 13-11: *Path dialect* — 327
Figure 13-12: *Calculation of* 5÷2 *in three dialects* — 331
Figure 13-13: *Bounding box derivation of Dominion* — 332
Figure 13-14: *Network and path derivations of Dominion* — 333
Figure 13-15: *Bounding box demonstration of unit-fraction bases* — 336
Figure 13-16: *Network demonstration of unit-fraction bases* — 337
Figure 13-17: *Path demonstration of unit-fraction bases* — 337

Figure 14-1: *Elements within Spencer Brown numbers* — 343
Figure 14-2: *Structure of Spencer Brown numbers* — 345
Figure 14-3: *Kauffman numbers in the form* — 348
Figure 14-4: *Kauffman two-boundary arithmetic* — 351
Figure 14-5: *Kauffman string arithmetic* — 353
Figure 14-6: *Kauffman string arithmetic with negative units* — 354
Figure 14-7: *Objects and operations in spatial algebra* — 357
Figure 14-8: *Cutting through symbolic distribution* — 360
Figure 14-9: *Models of multiplication* — 362
Figure 14-10: *Examples of two models of multiplication,* base-10 — 364

Figure 15-1: *Boundary arithmetic and boundary logic* — 373
Figure 15-2: *The principles of boundary arithmetic* — 378

Cast of Characters

Visionaries
Gregory Bateson
George Spencer Brown
John Horton Conway
Charles Sanders Peirce
Bertrand Russell & Alfred North Whitehead
Francisco Varela & Humberto Maturana
John Wheeler
Ludwig Wittgenstein
Stephen Wolfram

Voices
Stanislas Dehaene
Joseph Goguen
Louis Kauffman
Paul Lockhart
Alberto Martínez
Brian Rotman

Research and Perspective
John Barwise & John Etchemendy
Florian Cajori
Ronald Calinger
Philip Davis & Reuben Hersh
John Derbyshire
Mark Greaves
Georges Ifrah
Victor Katz
Donald Knuth
Alfred Korzybski
Imre Lakatos
George Lakoff & Rafael Núñez
Barry Mazur
Brian Cantwell Smith

Colleagues

Arthur Collings
Jack Engstrom
Thomas Etter
Fred Furtek
Robert Horn
Jeffrey James
David Keenan
Thomas McFarlane
Meredith Bricken Mills
Daniel Shapiro
Richard Shoup
William Winn

Special Thanks to

Colin Bricken
Ian Bricken
Julie Bricken
Jeffrey James
Louis Kauffman
Ted Nelson
Daniel Shapiro
Richard Shoup
John Walker

Historical Figures

	circa	*chapter*
Pythagoras	500 BCE	7
Democritus	400	7
Plato	400	7
Aristotle	350	7
Euclid	300 BCE	6
St. Augustine	400 CE	7
Fibonacci	1200	12
DaVinci	1500	7
Napier	1600	8, 12
Viète	1600	1
Galileo	1620	7
Descartes	1640	1
Leibniz	1700	1, 6, 12, 15
Joh. Bernoulli	1740	15
Euler	1760	2, 6, 15
Hume	1760	2
Lagrange	1800	1
Gauss	1840	1
Lobachevsky	1840	1
Bolyai	1860	1
Maxwell	1880	1
Kronecker	1880	8, 9
vonHelmholtz	1880	9
Dedekind	1880	9
Venn	1880	6
Frege	1900	1, 2, 6, 8, 9, 15
Peirce	1900	1, 2, 6, 9, 14
Peano	1900	1, 2, 9, 15
Hilbert	1900	1, 6
Brouwer	1920	9
Einstein	1920	1, 5

Quotes and Concepts

	chapter
Alain Badiou	1, 7
Gregory Bateson	1, 9
George Spencer Brown	1, 4, 5, 6, 7, 14, 15
P. Davis & R. Hersh	9
Richard Dedekind	2, 11

Stanislas Dehaene	1, 11, 15
John Derbyshire	15
Keith Devlin	3, 7, 12, 13
Leonard Euler	2, 6
Gottlob Frege	2, open-8
Joseph Goguen	1
Louis Kauffman	2, 3, 13, open-14
David Hilbert	1, 2
G. Lakoff & R. Núñez	1, 11
Mary Leng	11
Paul Lockhart	7, open-10, 11
Alberto Martínez	1, 2, open-12
H. Maturana & F. Varela	7
Thomas McFarlane	15
Charles Saunders Peirce	1, 2, 6, 14
Brian Rotman	1, 2, 7, 9, 11, open-13
Bertrand Russell	2, 6, 8
Ian Stewart	11, 15
Francisco Varela	open-15
A. Whitehead & B. Russell	5, 8
Ludwig Wittgenstein	11
Stephen Wolfram	1, open-4, 11, 14

Supporting Quotes and Concepts   *chapter*

Woody Allen	11
Aristotle	7
Zvi Artstein	1
Michael Atiyah	9
R. Augros & G. Stanciu	1
John Barrow	7
J. Barwise & J. Etchemendy	1, 6
Harry Belafonte	open-2
The Borg	10
Nicholas Bourbaki	1
L. Bunt, P. Jones & J. Bedient	6
Florian Cajori	3, 12
Ronald Calinger	11
Rudolf Carnap	1
Lewis Carroll	2

John Horton Conway	6
R. Courant & H. Robbins	7
Ubiratan D'Ambrosio	1
Leonardo DaVinci	7
Philip Davis	1
Paul Dirac	1
Albert Einstein	1, open-5
Jack Engstrom	8, 14
Richard Feynman	open-1
Vincente Garnica	2
Murray Gell-Mann	1
Henning Genz	7
Timothy Gowers	4
Alexander Grothendieck	6
Yuval Harari	7
Godfrey Hardy	1
S. Hawking & R. Penrose	1
Stephen Hawking	9
Richard Heck	9
Verina Huber-Dyson	11
Georges Ifrah	2, open-3
Morris Kline	15
Leopold Kronecker	9
Imre Lakatos	1
Jaron Lanier	1
J. Larkin & H. Simon	6
Gottfried Leibniz	6
John Littlewood	13
Danielle Macbeth	1
Ernst Mach	open-9
Barry Mazur	open-6
Joseph Mazur	11
Henri Matisse	1
Randall Munroe (xkcd)	11
Otto Neugebauer	11
Roger Penrose	1
Blaise Pascal	open-preface
Plato	7
Michael Potter	2, 9
Rudy Rucker	7

Jean-Paul Sartre	open-7
Herbert Simon	6
Brian Cantwell Smith	9
Irving Stein	9
F. Varela, E. Thompson & E. Rosch	1
John Wheeler	11
Alfred North Whitehead	6
Anthony Wilden	7
Hermann Weyl	open-11

Mentioned

Vladimir Arnold	6
George Boolos	2
Marilyn Burns	3
E. Gray & D. Tall	6
Mark Johnson	1
Alfred Korzybski	9
A. Michaelson & E. Morley	1
Rafael Robinson	1, 15
Vladimir Sazonov	11
Denise Schmandt-Besserat	1, 2
Daniel Shapiro	14
R. Vithal & O. Skovmose	1
William Winn	14

James Algebra

Chapters 5 through 13 in this Volume, as well as much of the content in Volumes II and III, explore James algebra. This boundary algebra was first developed between 1991 and 1993 by Jeffrey James and the author at the University of Washington Human Interface Technology Lab. The result of Jeffrey's research was published as his University of Washington 1993 Masters of Science in Engineering thesis, *A Calculus of Number Based on Spatial Forms*. James algebra is otherwise unpublished.

The abstract from Jeffrey's thesis provides an excellent summary of the original conception of James algebra.

> A calculus for writing and transforming numbers is defined. The calculus is based on a representational and computational paradigm, called *boundary mathematics*, in which representation consists of making distinctions out of the void. The calculus uses three boundary objects to create numbers and covers complex numbers and basic transcendentals. These same objects compose into operations on these numbers. Expressions transform using three spatial match and substitute rules that work in parallel across expressions. From the calculus emerge generalized forms of cardinality and inverse that apply identically to addition and multiplication. An imaginary form in the calculus expresses numbers in phase space, creating complex numbers. The calculus attempts to represent computational constraints explicitly, thereby improving our ability to design computational machinery and mathematical interfaces. Applications of the calculus to computational and educational domains are discussed.

Preface

*The last thing one knows when writing a book is
what to put first.*
— *Blaise Pascal (1670) Pensées*

Perhaps it's not particularly good form to begin with a confession, however: *This volume is a preamble.* My original intention was to explore **iconic logic**. I've elected to write first about arithmetic because not many people are familiar with symbolic logic. Boundary logic might be of interest to only a few and would risk putting the perspectives of iconic mathematics out of reach for many, particularly mathematics educators who rarely see logic within their standard curriculum. Arithmetic however no one escapes, not even preschoolers.

Formal logic has been cherished by Western culture as *the way that the mind works*. Supposedly, critical thinking is built upon logical, rational, formal thinking. Turns out that formal symbol systems have absolutely no correspondence to the way our minds actually work. Human dialogue and human values thoroughly embrace ambiguity, allusion, uniqueness, invention, emotion, multiple entendre, and at times, lawlessness. Studying logic is an arduous excursion into successive abandonment of the biological evolution of human cognition. Few understand how logic arises from nothing, or why we see binary values within an essentially unary form, or how imposed rigor completely avoids temporal and causal and sensual and social and situated information. Given the fundamental role that logic has played in intellectual history, yes our topic should be postsymbolic unary logic. I confess to taking the road more traveled.

George Spencer Brown's seminal work on iconic mathematics, *Laws of Form*, succinctly describes the structural foundations of mathematical thought. This volume is permeated with Spencer Brown's thinking and mathematical wisdom. His text is notorious. The academic analysis, enthusiasm, controversy and rejection of Spencer Brown's work is widely based on a severe misunderstanding that *Laws of Form* describes conventional logic, which it does not. The text becomes much more controversial when it is taken for what it actually is: a postsymbolic foundation for rigorous thinking.

To understand the foundation of mathematics
it is necessary to abandon
the symbolic representation of mathematics.

I've spent over three decades working professionally with boundary logic, yet this volume explores boundary arithmetic. Well, these volumes. Here's a second confession: in order to finish this volume, I had to keep spinning off chapters into secondary storage. The first to go into a separate container was *The Story of J*, which is [<( )>] when written as a James boundary form. J stands out whenever we do a particular kind of numeric thinking, one that postulates the existence of −1. The form of J exposes the consequences that follow when we imagine the possibility of bipolar units.

With J in its own volume, protecting if you will the rest of numerics from its influence, the content that remains weaves together two themes. There's *what happens* to common symbolic math when it is expressed within a postsymbolic perspective. And there's, um, *how we might think about what happens*. There is a *What!* theme and a *Whaaat?* theme. Both got too big, so the beast split again. Volume I explores the iconic form of arithmetic. Volume II explores the interface between iconic arithmetic and some of our current models of and ideas about mathematics.

Volume I begins with marks drawn in sand, with pebbles held in hand, with notches carved in bone, with unity standing alone. The tally system has been around longer than civilization. Its one-to-one correspondence anchors the birth of mathematical thought. **Unit-ensemble arithmetic** provides the first step toward calculation. Very early in our civilization, humans living in city-states put tallies into groups. These groupings eventually became digits. Throughout most of recorded history counting to determine *how* *many* was a specialist skill. A bucket of beans did not elicit a desire to know how many beans. Numbers were magical icons, only recently do we count. Symbolic arithmetic focuses our attention on the beans, postsymbolic arithmetic focuses on the bucket.

Strangely, the accumulation of tallies also gives birth to creatures that are not numbers. *Non-numeric* forms make an early and unavoidable appearance. We are left to contend with the realization that numeric form itself is a subjective imposition upon a broader terrain. Mathematics is about pattern, not number. Volume I builds arithmetic from *a single pattern*, that of **containment**. The non-numeric creatures have been exiled to Volume III.

OK, a third and final confession: it is with trepidation that I comment upon technical masterpieces in the philosophy of mathematics since I'm not a mathematician, have practiced very little abstract math, and have taught even less. I'm a computer scientist. Computer Science is a sister field to mathematics. The structural content it addresses creates a substantively different world-view about symbolic expression and about mathematical abstraction. Volume II particularly stretches into content that I've studied for decades and I still struggle to understand. How, for example, is it possible to separate form from function? How can rigor be independent of reality? How can information be context-free? How can we believe that thinking does not incorporate our senses? How can we pretend that there is a Platonic virtual reality that only our minds can access?

How can any human say that a computer has human characteristics? Computation is quite antithetical to organic existence. Neither does computation trespass into the purely imaginary realms of the infinite. Just like us, computing is embodied, but in a silicon housing that is far from biological. An algorithm has no access to the stuff that dreams are made of. It cannot think, it cannot make distinctions. The Turing test is a direct measure of human gullibility.

I began this project about a decade ago (in the ohohs), in a quite technical vein, wondering whether or not the axiomatic method developed for formal symbol systems could be applied to the simplest tallies used by humanity for thousands of years. Some of this work turned into Chapters 2, 3 and 4. But the motivation arose from a project a decade earlier (the 1990s if you are keeping track). I was working at Interval Research Corporation, Paul Allen's Silicon Valley research lab, with my long-time intellectual companion, Richard Shoup, and a team of inventive scientists including Tom Etter, Fred Furtek and Andrew Singer. Jeffrey James was also on the team for several years, contributing significantly to proof of principle. The *Natural Computing Project* was tasked with this challenge: if you could redesign computing from the ground up, without any consideration for what already exists, what would you build? Dick and I were both theoretical computer scientists interested in foundations, and we both believed that software languages were terribly impoverished, not because they failed to get a CPU to jump through hoops, but because they were built upon baroque mathematical presumptions. Natural Computing developed several alternative mathematics for computation including Shoup's Imaginary Booleans, Etter's Link Theory, Furtek's Torics Dynamic Constraints, and a diversity of supporting FPGA hardware architectures.

Together, Dick and I had both studied George Spencer Brown's seminal work a decade earlier. We understood Spencer Brown's iconic forms not as a path to Eastern philosophy, but as a tool for designing and writing better software languages and for building better computational hardware. During the 80s I met Professor Louis Kauffman. Lou's work conceptually extends *Laws of Form* and I've studied everything he has written. He inspired me to explore rigor creatively; his influence also permeates the mathematical content of this volume. Lou's development of iconic forms for iterated functions, for continued fractions, and for anti-boundaries such as)(, which he calls *extainers*, are excellent extensions of boundary thinking.

I came across *Laws of Form* a decade even earlier (the 1970s), guided by Stewart Brand's review and Heinz vonFoerster's commentary in The Whole Earth Catalog. At that time I was building a home in the forests of Hawaii, and had plenty of time to think about abstraction as I hauled lumber up a hill and pounded 200,000 nails to hold it all together. The unary logic in *Laws of Form* abandons Truth to Existence, a position that seemed quite reasonable to someone who was living in a forest. One day my mother showed up with a copy of *Laws of Form*, saying that she understood nothing inside, but it had literally jumped off the bookstore shelf as she passed by and landed at her feet. She thought it looked like something I might find interesting. Turns out, I'd been reflecting upon Crossing and Calling without the text for over a year. And I knew not to ignore Jungian synchronicity.

Before becoming a software designer, a decade before those other decades, I was first a teacher, and that's how I earned a living in Hawaii. The very first version of this volume turned into an extended rant about the state of math education in the USA. I was looking for better ways to construct respectful learning environments for growing children and had hoped that taking some of the cruft out of algebra might help. After a few hundred pages of analysis and criticism and ranting, it became obvious that math education is just too easy a target; it is too deeply flawed to justify the effort and the sincerity of a book. Criticism of educational practice implied that I should have been interested in helping to improve educational environments, and that was no longer the case. I had discovered a decade earlier still that *learning* is not within the charter of educational bureaucracy. A school is a place where three or more younger people meet with a state certified older person during specified daylight hours. That's it. I lost interest in schooling just like the Department of Education had lost interest in learning.

I spent the 80s learning CS and AI and ML and VR and UI and CAD and how to talk in acronyms, while becoming proficient in the design of computer languages. I viewed every formal discipline through the lens of boundary mathematics, bouncing ideas and perspectives off of several of the luminaries in Silicon Valley and every one of my professors at Stanford. A dear friend Daniel Shapiro shared many of these avenues of exploration, and both of us edited what the other wrote. My dissertation in part compared the error affordance of conventional and boundary notations for seventh grade algebra students. Net result is that algebra errors made by novices are contextual rather than symbolic, afforded rather than misguided.

We had better unwind. Experience in education led to interest in experiential learning which led to living in a forest which led to embodied math which led to Spencer Brown which led to a Ph.D. about symbolic thinking which led to our group at Interval, protected from the vagaries of underfunded institutions, trying to improve the foundations of computational mathematics. One result might be unexpected: our culture's time-honored and universally accepted place-value number system is not the only way to count, it is not necessarily the best way to do arithmetic, it is not even well designed.

Iconic math is rigorous thinking that looks and feels like what it is intended to mean. Postsymbolic thought is embodied experience. Our topic for the moment then is the deconstruction of common arithmetic based on the formal principles first developed by Spencer Brown, with the American philosopher Charles S. Peirce laying the groundwork at the turn of the twentieth century, and with our nomadic ancestors over 30,000 years ago providing tallies as the original substance from which numbers sprang.

Our ongoing cultural shift from text to imagery, from linear to parallel thinking, from encoded to experiential communication, from reading to watching to participating, suggests a shift from symbolic to iconic mathematics, just the kind of thing Spencer Brown has brilliantly delineated. It was a recent personal discovery that most influenced the evolution of this volume. I learned that marketing modern non-fiction books needs first a documented audience, and that in turn requires hyperactivity (for me at least) in the blogosphere and on Facebook and Twitter. Then I began to run into an embarrassing resistance from the written words. I was quite unable to write for a non-technical audience. In fact, it seemed as though I was unable to write for any kind of audience at all, much less to solicit and cajole that audience into existence. Only after resolutely resigning to my own inadequacies did the writing project become entertaining again.

So, here's a bonus confession: most of what is written recounts a conversation I have with myself. Credible fiction is too easy, I much prefer incredible non-fiction. Getting computers to run in unary logic, to abandon the concepts of True and False in favor of present and absent, to protest against models of cognition based upon symbol manipulation was my cup of tea. Arithmetic, something that everyone must deal with, like logic, is in great disrepair. There has been little of substance to improve our understanding of number One and number Two for over a century. If you have math anxiety, if you just don't get it, *it is not your fault.* Mathematics is to blame. Math teachers are asking us to play John Philip Sousa with a broken piccolo.

There are many ways to add 432 to 281. One is to realize that we have in our pockets an exact addition device, one constructed specifically for the task at hand, and all we need to do is to push some buttons. Before digital convergence, this device was called a phone. Another way to add is to incorporate cultural context and not care about exactness. We are putting a bit more than four buckets of a hundred together with a bit less than three buckets of a hundred, about 700 total. Another method is to realize that our culture is constructed so as to protect us from this type of onerous task, and to let a professional behind the counter such as a clerk or a teller do it. Another method is to be a cultural barbarian, a dinosaur. This type of addition is based on an absurd premise: use a paper and pencil and memorized algorithms to seek an exact symbolic number that has no meaning and no context for interpretation, while using a brain that has no evolutionary capacity to do so, and while ignoring our ubiquitous modern computing tools.

Writing words is much easier than implementing software. Words serve as fiction relative to the rigor imposed by automated computation. Dare I add another thread? The actual content of the numerics and logic and programming and education and even writing is focused on *postsymbolic methods of thinking*. What if arithmetic were pictures rather than symbols? Iconic rather than symbolic? Tactile rather than cognitive? Apparent rather than encrypted? Experiential rather than imagined? Concretely finite rather than abstractly infinite? What if numbers were to illustrate rather than encode? Pre-symbolic arithmetic was spatial, visual, tactile, and embodied prior to the symbolic reformation of the last century. The totalitarian dictate imposed by the Laws of Algebra has flattened space and touch and perspective and, yes, intelligence into rows of squiggles. Iconic arithmetic is intended to return life to numbers.

So we arrive at the pregnant question: what *should* arithmetic feel like in this century? Exploring and playing with and getting the feel of iconic arithmetic can be astonishingly familiar, it is how arithmetic was before universal schooling sucked the life out of it. If we replace abstraction by embodiment, will mathematics return to Earth? Might we then as a civilization become more aware of and more careful with the only world we have? Is mathematics contributing to the virtualization and then the trivialization of physical reality? Are numbers denizens of cyberspace, sirens that lure us away from stewardship and into fantasy? What will ecological arithmetic look like?

Here's where we stand: three volumes of numerics. The first attempting to save children from symbolic abuse, the second on the formal axiomatics and philosophy of boundary arithmetic, and the third sharing a most exciting feature of the imaginary realm. These three volumes then provide familiarity before wading into unary logic, before the volumes that will explore how to be both rigorous and rational by forgetting what doesn't matter, by ignoring the concept of False as completely irrelevant, by reconnecting rationality to sensory presence, by integrating physical intuition with cognitive visualization, and by treating dichotomy itself as an illusion.

The present goal is to develop an intuitive arithmetic that is both rigorous and embodied, to restore arithmetic to the simplicity of its Babylonian origins as piles of pebbles. The motivation is to provide both school children and adults with an alternative to math-by-symbol-manipulation, an alternative that maintains the essence and rigor of mathematics while also accommodating the essence and sensibility of humanity.

The website iconicmath.com is mentioned frequently as a resource for videos and correspondence. It preceded and served as an outline for what you may choose to read next. The Cast of Characters are both colleagues I have learned from and authors I have studied. All are mentioned within the text. This volume does not consolidate what is known nor reflect the views of others. It is an exploration. Welcome to a postsymbolic playground.

> william bricken
> Snohomish Washington, July 16, 2018

Iconic Arithmetic
Volume I

Chapter 1

Context

You can recognize truth by its beauty and simplicity.[1]
— *Richard Feynman (1985)*

We are about to explore **boundary mathematics**, a completely different perspective on the common mathematical tools that we use daily. Boundary math is built from *icons rather than symbols*. **Iconic math** is embodied rather than abstract.

Boundary arithmetic relies upon the single physical relationship of *containment* to express the elementary ideas of arithmetic. In this volume, we will be exploring two types of boundary arithmetic. **Depth-value unit ensembles** unify the way we write numbers with the way we add and multiply numbers. **James algebra** defines the common concepts and operations of arithmetic, (count, +, −, ×, ÷, ^, log, √), as different ways of arranging containers.

The objective is to learn more about conventional arithmetic as it unfolds within the unconventional conceptual system of nested containers.

Chapter 1

Modern mathematics is an intellectual infrastructure for solving problems in science, commerce, engineering and technology. It is a vital tool for our civilization. The magnificent edifice of *advanced math* is not in question, since the few who practice advanced math have been extensively trained in navigating the treacherous waters and the precipitous chasms between formal concept and informal communication. Presumably we teach math to everyone because it helps with rational thinking, with formulating a scientific world view, with creating a better world. Since math is a way of thinking, it makes sense to include elements of how humans think into the structure of mathematics, without compromising the formality that distinguishes math from the other disciplines. But **symbolic mathematics**, as currently taught and practiced, is disconnected from human evolution, from human learning, from human psychology and from our natural human capabilities.

Different types of math engender different types of thinking. This volume is also about *a new way of thinking*. Arithmetic serves only as the context for a broader idea. As a small step, if we can come to understand a different kind of math, one that is more natural and more visceral, then we might at least be armed with the knowledge that the math taught in schools is a **design choice**, not a necessity. Although it is possible to read iconic boundaries as common arithmetic, it is also possible to explore iconic modeling without ever considering it to be about numbers or arithmetic or algebra. Boundary math is about **cognitive distinction**.

> *The objective is to learn more about how we think by exploring the formal structure of distinctions.*

If there is a single guiding light it is to recapture the simpler ways of doing arithmetic that evolved in human cultures over millennia, to return to a mathematics that makes sense because it is sensual.

1.1 Communication

A **symbol** is an encoded chunk of information. Symbols are communication tools that have an *arbitrary* representation. Because symbols are encoded, we have to memorize and to recall the patterns that they weave. That's why we have to *teach* children to read and to count. But not to walk, to talk, to see or to think. Symbols specialize in abstract ideas like freedom and ethics. Symbolic description is easily standardized. Symbols also impose a substantial **cognitive load**. It's not a good idea to try to solve a symbolic algebra problem while playing tennis.

symbol

HOUSE

An **icon** *looks like what it means*. It bears a structural resemblance to the ideas it is intended to convey. Icons are communication tools that permit our senses to make the connection between image and meaning. An icon's structure reflects its intention. Iconic arithmetic has a look and feel that connects to our bodies as well as to our minds.

icon

The representation of the concept *five* should look like the icon ///// rather than the symbol 5. The concept of *nothing* should not have a representation. The symbol 0 looks like something, it visually contradicts its own meaning. "0" is not nothing.

image

The foundation of iconic arithmetic is the Additive Principle.

Additive Principle
A sum looks like the collection of its parts.

//   /// = /////

The principle is physical, based on appearance as well as concept. It is the definition of addition that has been with us since the beginning of civilization. Adding is putting things together. Putting together does not change what things look like. The Principle of Multiplication is that every part of one *touches* every part of another. Multiplication is complete connectivity.

Multiplicative Principle
Every part of one contacts every part of another.

Chapter 1

Like any math, iconic math is rigorous. It gives the same results as symbolic math while also maintaining a connection to concrete images and to familiar experiences. As iconic languages, then, James algebra and depth-value unit ensembles have the additional (some would say non-mathematical) requirements of being both **sensual** and **concrete**. The concept called contains can be thought, seen and felt.

Three Volumes

arithmetic is putting stuff into boxes

In this volume, we'll begin with the arithmetic of tallies, simple marks that generate numbers. We'll see how to add and to multiply tallies, in the process generating unit-ensemble arithmetic. We'll use depth-value notation to create parens arithmetic, a typographical yet iconic form. Then we'll expand parens into two and three dimensional spatial dialects and give them dynamics. In Chapter 5 we next explore the structure of James algebra. We'll build common math out of three patterns of containment. Two of these patterns show us what can freely be discarded by calling upon the powerful strategy of **void-equivalence**. We'll look at some examples of experiential dialects of James algebra, and we'll conclude the volume by quickly exploring some other boundary math systems for arithmetic.

Volume II examines the relationship between James algebra and the concepts that currently define formal mathematics. We'll compare boundary mathematics to Frege, Peano, Robinson and other symbolic definitions of number and then make the case for postsymbolic thinking.

Volume III contains a surprise, a new imaginary number. Conventional analysis focuses on imaginary numbers with their characteristic unit i. At the core of James algebra is a new imaginary unit, J. As an anchor to familiar concepts, J can be interpreted as the neglected logarithm of -1. The multiplicative imaginary i is a *composite* of the simpler **additive imaginary** J. In Volume III we'll also explore non-numeric infinite and indeterminate James forms.

Finally a gentle if risky reminder. We will be exploring both the **formal structure** and the conceptual development of common numbers. Our task is to understand what numbers are and how they work by describing them in an unfamiliar foreign language. That language has *making a distinction* (observing a difference) as its primitive operation. Standing back, we'll see that the entire mechanism of arithmetic and algebra can be expressed by three permissions to change structure. Two rules permit structure to be discarded, one permits rearrangement. Standing close, we'll see that the fine-grain structure of arithmetic is incredibly simple.

Once you get comfortable with the idea that structural transformation is independent of interpretation, your eyes and fingers can take over for your brain. Learning to abandon the conceptual and notational mire that we have been taught as common arithmetic no doubt will be a challenge. The most dominant obstacle is our natural propensity to impose previously learned complexity upon the simplicity of void-based reasoning. We'll abandon the use of symbolic expressions in order to see more clearly. We will be looking toward what Jaron Lanier calls *postsymbolic communication*.[2] An irony is that symbolic mathematics is only about a century old: more accurately we are looking backward, to the way that mathematics and particularly arithmetic has been carried out for thousands of years.

The objective is to be able to do pre-college math with our eyes and with our fingers.

1.2 History

Throughout history, humans have used the abacus, the counting table, the knotted rope, the tally stick and the parts of our bodies to assist with the tasks of arithmetic.[3] We now have superb digital tools that have replaced physical tools such as pencil and paper. It is time that we end the pretense that people should know the algorithms

Chinese abacus

Chapter 1

European counting table circa 1550

used to calculate with numbers. Addition and multiplication are no longer mental skills, they are buttons on an electronic device.

The brain is the wrong tool to use for calculating.

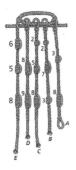

Incan quipu

For thousands of years, counting in Western cultures (ancient Greek and Roman societies for example) was done by one-to-one correspondence. The introduction of Hindu-Arabic numerals around 800 CE gave numbers their current meaning.[4] Then for another thousand years, until the nineteenth century, counting (and accounting) was done by professional counters, on a counting table, in a counting house. These houses became so popular that they eventually turned into banks.

Gottfried Leibniz 1646–1716

Classical geometry is a premiere example of an iconic system. The objects of interest, squares and circles and angles, look just like squares and circles and angles. Leibniz, Descartes and Viète converted equations consisting mostly of abbreviated words into symbolic systems that later came into wider use during the seventeenth century. In the mid-nineteenth century the discovery of non-Euclidean systems of geometry contributed to a loss of trust in human mathematical intuition, particularly in our spatial and visual senses. Then we got carried away. Here's mathematician Philip Davis:

For two centuries mathematics has had harsh words to say about visual evidence. The French mathematicians around the time of Lagrange got rid of visual arguments in favor of the purely verbal-logical (analytic) arguments that they thought more secure.[5]

Joseph-Loius Lagrange 1736–1813

The math that we teach in schools lurched into the twentieth century on the back of this crisis in confidence. Those in the mathematical community who wondered about the rigor of mathematics discovered, after thousands of years, that they did not really understand arithmetic or geometry or badly behaving functions or even rigorous thinking. And so the community adopted a radical plan to put mathematics on a firm foundation. The hot idea was **symbolic formalization**, representing concepts using encoded symbols that bear no resemblance to the concepts they identify. Mathematical concepts were to be conveyed using strings of typographic characters rather than using pictures and physical objects and overt behavior. Meaning was to be embedded into the sequential patterns of meaningless squiggles. The disconnection of form and meaning seemed reasonable since, from a Platonic perspective, abstract concepts do not dwell in physical reality. If we cannot point at a concept such as *all numbers* then we surely cannot illustrate it. Any squiggle, perhaps $\forall n$, will do.

David Hilbert popularized among mathematicians the idea that math can be made formal and thus certain by removing all meaning from mathematical symbols, thereby relying solely on the structural relations between symbols. Foundational logician Rudolf Carnap:

David Hilbert 1862–1943

> A theory, a rule, a definition, or the like is to be called formal when no reference is made in it either to the meaning of the symbols (e.g. the words) or to the sense of the expressions (e.g. the sentences), but simply and solely to the kinds and orders of the symbols from which the expressions are constructed.[6]

Chapter 1

In an ironic twist, Hilbert then argued that arithmetic carries its own meaning.

> In elementary domains of arithmetic...there is that complete certainty in our considerations. Here we get by without axioms, and the inferences have a character of the tangibly certain.

Hilbert continued that the essential difference between the representation of a number and the concept of a number is that

> *the object doing the representing contains the essential properties of the object to be represented.*[7]

Brahmi digits circa 800 CE

Hilbert was talking specifically about putting iconic strokes together to yield sums (e.g // + /// = /////), the unit-ensemble model described in Chapter 2. Thus we are in complete agreement with Hilbert: numbers are essentially iconic.

Symbolic math was *invented*, along with the horrible design idea that math should be done with our minds and our memories, rather than with our eyes, our bodies and our physical tools. The unifying theme was that every mathematical object is a *set*. The goal was to endow math with **purity**, to collapse all of mathematics into one grand scheme. Assumed invariant patterns called **axioms** defined how to think rigorously. Nicholas Bourbaki:

Bourbaki School 1935

> The internal evolution of mathematical science has, in spite of appearance, brought about a closer unity among its different parts...and which has led to what is generally known as the "axiomatic method."[8]

The great success of the entirely symbolic approach, followed closely by the rise of the use of streams of ones and zeros in digital computers, has led to the expression of mathematics almost exclusively in symbolic string languages. Unfortunately, in the rush to make math symbolic, the early twentieth century founders seem to have forgotten that people care about *understanding* and *using* math, not about an esoteric collection of structural rules that cast common arithmetic into conformity with a maze of symbolic concepts delineated by group theory built on top of set theory built on top of predicate logic. Logicians Barwise and Etchemendy observe:

> Despite the obvious importance of visual images in human cognitive activities, visual representation remains a second-class citizen in both the theory and practice of mathematics.[9]

Belief

Mathematics is a human endeavor, obviously. Mathematics is replete with metaphysical concepts (also called *beliefs*) such as completed infinity, physical continuity, Platonic virtual reality, universal and eternal truths, constructions so grand that they are beyond the scope of time and space, extra-human origins of human ideas, calculations that may not halt, and formal objects that exist but cannot be identified. These belief systems are not necessary for a formal description of nor for an informal understanding of arithmetic.[10]

Perhaps there is not one grand system that unifies mathematics. As scholars studying ethnomathematics have affirmed, mathematical thought exhibits organic diversity.[11] Here is computer scientist Joseph Goguen's attempt to disperse the totalitarian attitude that mathematical thinking is uniform and universal:

Joseph Goguen
1941–2006

> Notation is only the surface reflection of these deeper, essentially social struggles. But perhaps it is time we realize that no metanarrative can be demonstrably superior to all others for all purposes, and that, in this sense, we live in many different worlds, rather than in just one world.[12]

A premiere example of a physically dysfunctional symbolic concept is the associative law of addition.

$$(a + b) + c = a + (b + c)$$

The parentheses used to assert this law create visual interference that undermines the meaning of addition, that of **putting together**. Putting things together two-at-a-time is not a property of addition. The associative law specifies a *method* to achieve addition, one that is not particularly efficient. The sequential two-at-a-time strategy might be because there are two sides to a textual plus sign; it might be because addition tables are constructed to add only two numbers at a time; it might be because relations are usually binary; it might be because group theory incorporates right- and left-side rules; or it might be a hangover from counting by adding one at a time. But here's what we can learn from children. *To add many things, put them together. It doesn't matter how.*

Vision

The earliest founders of rigorous mathematical systems used diagrams and visual thinking extensively. Venn and Frege and Peirce all developed functional visual mathematical notations. They understood that thinking unites imagery with structure. According to C. S. Peirce,

> All necessary reasoning is diagrammatic. ... all reasoning depends directly or indirectly upon diagrams.[13]

Many rigorous iconic calculi have recently been studied; these include fractals, cellular automata, particle diffusion, knot theory, and the paradigm of modern algebra, category theory. Philosopher Danielle Macbeth:

> At least for the case of mathematical concepts, then, we can say exactly what meaning is: it is nothing more and nothing less than what is exhibited in... diagrams and expressions that directly display the senses of concept words, senses within which are contained everything necessary for a correct inference.[14]

Mathematics is necessarily sensual, so let's pardon it from banishment into symbolic disembodiment. Let's ground arithmetic in the tangible Earth rather than in imaginary realms of conceptual abstraction and let's require that it recognize reality as we humans experience it.

1.3 Boundary Forms

Boundary forms are *configurations of nested boundaries*. Boundaries both separate and connect. It is convenient to consider these boundaries as **containers** that have an inside and an outside. In this volume, we'll turn the concept of containment into a comprehensive mathematical tool.

separated

connected

In James algebra for example *everything* is a container. Empty containers are units. There are two basic forms: ( ) and [ ]. They cancel each other out. When we wish to connect to arithmetic, ( ) is One and [ ] is a type of Infinity. When we wish to think about cognitive distinctions, we might say that ( ) is *what is* and [ ] is *what could be*. When we wish to eschew interpretation, the name of ( ) is **Round** and the name of [ ] is **Square**. These delimiters and their names are quite arbitrary, dictated by the available keys on a typewriter, and having no relationship to the geometric shapes they may invoke. These delimiters represent generic containers. The only thing that a container does, regardless of shape or representation, is to hold things.

some containers

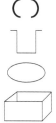

Chapter 1

A central innovative idea is to treat non-existence, *void*, with the respect it deserves. *Void* has no properties and does not exist, taking the concept of 0 with it into oblivion. An **empty container** is *void* inside. Empty containers are **units** precisely because they hold nothing inside.

an ensemble

A collection of units within a common container is an **ensemble**. We can interpret the *cardinality* of an ensemble as a **number**. Addition is eliminating the boundaries that distinguish separate ensembles. Multiplication is putting ensembles inside other ensembles. Addition eliminates distinctions while multiplication nests them.

James algebra includes a **generalized inverse**, < >, which is a *single* boundary concept that condenses the basic inverse operations of arithmetic. Negative numbers, subtraction, unit fractions, division, roots and logarithms are essentially the *same operation* applied in different contexts. Cognitively, < > can be associated with *reflection*, with turning structure back upon itself. It's name is **Angle** and alone it is nothing.

Containers

The concept of a container is tangible, not abstract. *All boundary forms can be constructed in physical space.* Containers are fundamentally different than symbols, containers have a physical presence. Containers also have an *inside*. Here's Lakoff and Núñez as they develop a theory about the cognitive origins of mathematics:

> The Container schema has three parts: an Interior, a Boundary, and an Exterior. This structure forms a gestalt, in the sense that the parts make no sense without the whole.[15]

Containers are so useful that they form the structural basis of

— computer circuitry and logic
— mathematical sets and functions

- web description languages like HTML and XML
- software design environments like InDesign
- storage closets
- loading docks
- suitcases and
- kitchens.[16]

A container distinguishes two spaces, its inside and its outside.[17] It is an object from the perspective of its outside **context**, and an operator upon its inside **content**. When we look at a container from the outside, it appears to be an independent object. When we are inside a container, it feels like an environment.[18] When we make a distinction, we are its container. It is *our viewing perspective* that determines whether a container is an object that can be manipulated or an environment that can be experienced or a concept that can be imagined.

object

environment

Boundaries invite participation. This feature is sufficiently important to elevate it to our first principle, the Principle of Participation.

Participation
How we look at a form determines what it is.

distinction

There is *only one relationship* between two containers, A contains B. Containment, together with the structure of boundaries, is sufficient to express completely the objects and operations of elementary arithmetic and algebra. This is what it means to *simplify* arithmetic: to describe it fully using one idea, the **distinction**, as represented by a container that distinguishes inside from outside.

A contains B

$(B)_A$

Containers find their cognitive expression as *categories*, which are firmly grounded in the sensory-motor systems of the brain. Lakoff and Núñez again:

> Propositional calculus (the simplest form of symbolic logic) is ultimately grounded in Container schemas in the visual system.[19]

From an abstract perspective, the name "contains" is arbitrary, a metaphor for visualization but not essential. We are exploring a system with one **binary relation**, R, that has the properties of being irreflexive, asymmetric, and intransitive. These properties are more or less the opposite of the properties of equals. In other contexts, R might be called parent, underneath, eats, causes, implies, and an unlimited number of other evocative names.

Distinction

Well, there is a refinement to be made. The idea of a "binary relation" comes from our conventional foundation of logic and sets. Relational concepts such as reflexive and symmetric and transitive are themselves derivative rather than definitional. We are beginning *prior to* logic, prior to the formal concept of a relation. The abstraction that generalizes the contains relation is **distinction**, or *difference*. Spencer Brown provides a definition:

distinction

> A distinction is drawn by arranging a boundary with separate sides so that a point on one side cannot reach the other side without crossing the boundary.[20]

Distinction is not numeric, it has no units of measurement and no scale of comparison. "Difference is not a quantity."[21] As such, distinction is not grounded in science, it is grounded in experience. Gregory Bateson continues:

Gregory Bateson
1904–1980

> A difference is an elementary idea. It is of the stuff of which minds are made.[22]

Our minds and our senses transact differences. Matisse: "I don't paint things. I only paint the difference between things."[23]

An algebra of distinctions is fundamentally different than both an algebra of measurements and an algebra of numbers. Actions in the world involve energy. We might say that they occur *on the outside* of the boundary of our bodies. In contrast, difference does not involve energy. Difference is not localizable, it cannot be assigned a place or a time. Difference is *on the inside*. When we compare a duck to a chicken, the difference does not reside within the duck, nor is it within the chicken, nor is it within the space that separates the two. "In a word, *a difference is an idea*."[24] Let's elevate Bateson's definition to our second principle, the Principle of Distinction.

difference is dimensionless

Distinction
Difference is an idea.

Our bodies then are the concrete realization of physical difference, while our thoughts are the abstract realization of virtual difference. Bateson provides clarity,

> The explanatory world of *substance* can invoke no differences and no ideas but only forces and impacts. And, per contra, the world of *form* and communication invokes no things, forces, or impacts but only differences and ideas.[25]

Without distinction, there is the **non-concept** *void*. This underlying page supports typographical characters. Similarly *void* is the substrate that supports distinction. Like the white space of the page, *void* is everywhere. Unlike this page, *void* has absolutely no properties of any kind. This means that there are no relationships between containers within the same space since there is no intervening medium to support a relationship. No points, no distances, no coordinates, no dimensions, no metrics. It is as if the characters printed on a page were each independent and unrelated to one another. All that they would share is the page or the line they are recorded upon. There is only containment. Our third principle:

Void
Void has no properties.

1.4 Embodiment

Our senses and our bodies are the interface between manifest and conceptual. Distinction is the ground of perception. When there is no distinction between inside and outside, we do not perceive a difference. The interface between our physical and our cognitive selves is a boundary that transacts only differences. The physicality of our body defines an obvious container of our self. Boundary math carries this physicality, via iconic representation, directly into the core of mathematical thinking. Philosopher Mark Johnson:

> The container schema's structural elements are "interior, boundary, exterior," its basic logic is "inside or outside," and its metaphorical projection gives structure to our conceptualizations of the visual field (things go in and out of sight), personal relationships (one gets in or out of a relationship), the logic of sets (sets contain their members), and so on.[26]

James algebra and unit ensembles are both **embodied**. The *concept* of a container is necessarily abstract. The representation of a container is physical, a tangible manifestation of the concept. The meaning of the representation is unequivocally an idea, an idea that can at any time be enacted. Lakoff and Núñez: "Ultimately, mathematical meaning is like everyday meaning. It is part of embodied cognition."[27]

nominalism
abstract objects do not exist

Mathematics and mathematical "truth" necessarily co-evolve, not only in a conceptual sense, but also in coordination with the physical evolution of our brains and our bodies. We come to understand mathematical ideas *through* our bodies. Bluntly, mathematics is "conceptualized by human beings using the brain's cognitive mechanisms."[28] Containers are intended to re-anchor mathematics to physical existence, to return the beast to its creators, to tame those aspects of the beast that we might encounter in daily life. We have evolved from mathematical Platonism to mathematical nominalism.

Rigor

Rigorous math requires that mathematical techniques be independent of our personal and cultural biases and beliefs. We want math to be extremely useful and completely reliable and also uncompromising in its neutrality. But does neutrality necessarily imply non-human? Must math be an abstract fantasy removed from human existence? Can math be both rigorous and organic?

In *Proofs and Refutations* Lakatos chronicles the evolution of a single formal idea, the structure of polyhedra. He concludes:

Imre Lakatos
1922–1974

> Mathematics, this product of human activity, 'alienates itself' from the human activity which has been producing it. It becomes a living, growing organism, that *acquires a certain autonomy* ... its own autonomous laws of growth, its own dialectic.[29]

It is time to reclaim mathematics as a human activity, to tame the creature we have unleashed. **Humane math** aligns with how our minds are known to work, with how the patterns of communication between people work, and with how physical existence embodies our knowledge. *Boundary math is both humane and rigorous*, the two objectives are not incompatible. The result is a much simpler iconic arithmetic that reflects how numbers evolved in human cognition prior to our recent truly exotic experiment with purely symbolic mathematical form.

Truth and Beauty

This exploration has deep parallels with the way that physicists describe reality. The basic technique of Physics is to derive simple mathematical models that condense disparate observations. Here we are exploring the basic nature of arithmetic by developing a simple model that condenses disparate algorithms and operations. Einstein:

Chapter 1

A theory is more impressive the greater the simplicity of its premises is, the more different kinds of things it relates, and the more extended is its area of applicability.[30]

Two different types of problem might arise. Further observation may introduce new data that is inconsistent with the model. In this case the model needs to be expanded, or perhaps abandoned altogether. Problems can also occur *within* the model itself. Perhaps a nasty division by zero might be encountered. An intractable infinity might be swept away through an inventive mathematical technique such as renormalization. Or the model might be abandoned altogether. Such was the case in the early nineteenth century when Lobachevsky and Bolyai (and Gauss before them) developed geometries that abandoned the parallel lines axiom of ancient Greek geometry.

It was Maxwell's electromagnetic equations in the late nineteenth century that first challenged the idea that physics was about perceived reality. His wave equations ignored both the how (i.e. the physics) and the what (i.e. the observable). Mathematician Zvi Artstein:

> The only justification for the existence of the waves was the facts that they provided a solution to the equations, solutions that have wavelike characteristics....He forwent the physical explanation based on known physical quantities. He published his equations, as if declaring this is physics, the physics is inherent in mathematics.[31]

$\nabla \cdot \mathbf{D} = \rho$
$\nabla \cdot \mathbf{B} = 0$
$\nabla \times \mathbf{E} = -\frac{\partial \mathbf{B}}{\partial t}$
$\nabla \times \mathbf{H} = \mathbf{J} + \frac{\partial \mathbf{D}}{\partial t}$

When Michaelson and Morley went looking for the aether that supported the wave propagation of light they found nothing. There was no physical substrate.

Maxwell's equations are beautifully symmetric. Beautiful models, those with lots of symmetry and regular structure, have been preferred historically over models that are messy. Scientists Augros and Stanciu: "Beauty is the primary standard for scientific truth."[32] The same is true of mathematics. Here's mathematician Godfrey Hardy:

> Beauty is the first test: there is no permanent place in the world for ugly mathematics.[33]

This despite the possibility that both Nature and mathematics may at their core be messy. To maintain mathematical beauty, new, never before observed phenomena might be postulated. Experiments are then designed to expose the potential new phenomena. Theory drives experiment. Often different theories diverge, to arrive at completely different *descriptions* of the same observations. In this case, attempts are made to merge the theories, or perhaps the theory with the greatest explanatory power is kept and the others tossed away. Physicist Murray Gell-Mann:

> Frequently a theorist will throw out a lot of data on the grounds that if they don't fit an elegant scheme, they are wrong.[34]

What is *true* is moot, since we can also be guided by what is *simplest*. The field of mathematics has recently abandoned the connection of Truth to Reality by defining its own truth as its own internal consistency. The concept of Truth itself can work against accurate description, as it does in the case of the wave/particle duality in quantum mechanics. Physicist Paul Dirac:

> It is more important to have beauty in one's equations than to have them fit an experiment.[35]

This attitude stems from a growing realization that modern physics is no longer connected to our perception of Nature. Mathematical physicist Roger Penrose:

> It is a common view among many of today's physicists that quantum mechanics provides us with *no* picture of 'reality' at all! The formalism of quantum mechanics, on this view, is to be taken as just that: a mathematical formalism.[36]

When physics becomes mathematics, our concepts of physical reality begin to conform to the values of mathematics. Elegance, symmetry, scope, these are how leading scientists view Nature. Here's Stephen Hawking:

> I don't demand that a theory correspond to reality because I don't know what it is. Reality is not a quality you can test with litmus paper. All I'm concerned with is that the theory should predict the results of measurements.[37]

And we know from quantum mechanics that the results of measurements, what we see, depend upon how we elect to look. Reality cannot escape its sensory basis.

1.5 Humane Mathematics

If metaphysical Truth and physical reality are no longer the criteria for knowledge, are there multiple ways to describe the concept of number? Which are preferable? Specifically, which description of numbers, a beautiful concise one or a messy complex one, is preferable? Are any particular models of mathematical patterns more true than others? Is the way we write arithmetic fully separate from what we are attempting to express? Is the barrier between representation and meaning an impermeable wall, or is it perhaps a bridge, or is it a fantasy of convenience that supports meta-analysis at the cost of comprehension?

In the realm of physics, experiment and observation often help to guide the choice of language for a theory. In elementary mathematics, the ability of children to learn simple concepts can be taken as experimental evidence. A mathematical concept is not simple if kids do not understand it. Our current observational data is that after studying arithmetic for a dozen or so years, about two out of every three children and young adults in the U.S. do not understand such elementary ideas as negative numbers, fractions and the rule of distribution that connects addition and multiplication. These three concepts account for the vast majority of student errors in math classes.[38] Might the wisdom of youth be telling us that perhaps we have the wrong model of arithmetic itself? Cognitive scientist Stanislas Dehaene:

> When we think about numbers, or do arithmetic, we do not rely solely on a purified, ethereal, abstract concept of number. Our brain immediately links the abstract number to concrete notions of size, location and time.[39]

We have a choice to continue to support a mathematical and symbolic perspective that denies human experience and the reality of our physical presence in a physical environment, or to encourage students to be aware and responsible within the world of physicality by teaching mathematics as a pragmatic tool rather than as an abstract, cognitive skill.

The vision is to return to **native mathematics**, prior to the conversion of the visceral sensate numbers of our cultural and physical evolution into the meaningless symbolic squiggles of the twentieth century. Symbolic math may have been completely appropriate during the transitional period from about 1910 to 1980, prior to the computer revolution. Now it is time to give symbolic processing to its rightful owners, those microscopic arrays of transistors and wires that constitute digital hardware. It is time to return mathematical concept, understanding, intuition, and just plain common sense to their rightful owners, flesh-and-blood humans. We can reserve symbolic math for professionals, and for students who declare their interest in the field of mathematics as a college major. We can maintain **formalism**, the essence of modern mathematics, within precollege math without loosing our humanity by embodying it in utility. We do not need to pretend that math exists in some abstract virtual world. We can own math as a human invention, a tool of thought.

Arabic digits circa ~800 CE

Chapter 1

We do not need math classes for non-mathematicians.

Any subject can call upon the tools provided by a mathematical perspective, when needed, when appropriate, and when useful. All it takes is recognition of the obvious: symbolic math is not humane math. Ask any school child.

Simplicity

What is necessary first is to show that non-symbolic math does exist, that we can anchor the formal concepts of arithmetic within common sense and within common sensation. Here's Stephen Wolfram:

> Like most other fields of human enquiry mathematics has tended to define itself to be concerned with just those questions that its methods can successfully address...the vast majority of mathematics practiced today still seems to follow remarkably closely the traditions of arithmetic and geometry that already existed even in Babylonian times.[40]

There is an organic legitimacy, a human context, from which mathematics arose. But we became enamored by the symbolic process, without adequately recognizing the cost. Historian of mathematics Alberto Martínez comments:

> Even when elementary mathematical rules are designed to represent plain manipulations of things, they might still be used to construct symbolic statements that do not correspond to anything that can be exhibited palpably.... Throughout history, mathematicians realized that by adopting diverse and particular empirical explanations to justify specific symbolic operations, mathematics acquired a semblance of arbitrariness and inconsistency. Thus they came to cast aside empirical explanations as mere illustrations and applications, and not as justifications for mathematical rules.[41]

The concept of truth has migrated from experience to symbolic structure. Mathematics abandoned the physical in favor of its own internal structure. In the first half of the twentieth century, however, it became clear that mathematical structure had neither global truth nor consistency to offer. Martínez again:

> Virtually nobody imputes to traditional algebraic methods any blame for instances where the symbolism generates unsatisfactory or bizarre results... virtually nobody says that maybe it is the algebra that is defective.[42]

Suzhou digits circa ~1200 CE

Professional mathematicians are trained to use powerful and exotic tools, and advanced math is such a tool in the hands of a professional mathematician. But when, for example, should an average driver know how to fix an electronic carburetor? When should a parent know how to prepare *quiche lorraine* for breakfast? When should a web surfer know how to write XML database search algorithms? When should a homeowner use the quadratic formula?

The *discipline* of math is built upon absolute abstraction. It is intentionally and adamantly disembodied (ignoring the human), ungrounded (ignoring the earth), and imaginary (ignoring reality). Philospher of mathematics Brian Rotman observes that mathematics itself is built upon

> A world of mathematical objects — numbers, points, lines, sets, functions, morphisms, spaces, and the like — that are held to exist prior to and independent of any talk, description, or discussion of them....the belief in objects "out there" — uncorrupted by the vagaries and uncertainty of history, culture, human choice, and the associated subjectivities that permeate discourse — is crucial and nonnegotiable.[43]

Chapter 1

*European digits
circa 1900 CE*

1
2
3
4

The concepts that constitute abstract math (symbolic manipulation, Platonic reality, infinity, zero) are *theological*. It is churches that manipulate sacred symbols as doctrine; it is religions that call upon an imaginary reality for miracles; it is only gods that are infinite. Math has established itself to be sacred, to be beyond humanity, and thus to be beyond common comprehension. Unlike religion, however, one's belief in math is not rewarded by absolution or by contentment or by virtue. Failure to believe in the whole numbers leads to extreme derision. A numeric atheist (a matheist) gets condemned not to hell but to ignorance. While common religion allows the needy the salvation of belief, common math simply equates need with incapability.

The **Doctrine of Abstraction** is unique to Western academic mathematical thinking. Street merchants in Brazil, money lenders in India, school children in Africa, grocery shoppers in an American supermarket, even statistical scientists measuring the popularity of a political candidate, none comprehend numbers as abstractions. Their real-world numbers are constantly grounded in application, in utility, in human dynamics. **Ethnomathematics**, the study of how people actually use math, sees the extreme abstraction embodied in Western thought as yet another mechanism of cultural imperialism.[44]

The mathematics of chaos theory, iterated functions and cellular automata each embody an anti-abstraction principle. Although computation in these fields can be deterministic, the only way we know what the next step will bring is to compute that step. It is not possible to develop a symbolic abstraction or condensation that models or that predicts what is next. There is nothing simpler than real-time unfolding.

Yes, it will probably be very embarrassing for our culture to admit how delusional it has been, but by now we should be accustomed to delusional error. No, the Earth is not the center of the universe. Those twinkles in the sky are not all stars. Indigenous peoples are not savages. Obviously all mammals have emotions. The universe is not made of matter, nor is it made of mathematics. Mathematics itself is the tool that allows us to explicitly ignore the deeper detail of our world in favor of a crisp abstract summary that we can see, handle and use.

1.6 Remarks

Symbolic math is a big part of the problem of math education; another big part is our cultural belief that all people should learn the details of an esoteric discipline at the cost of their own self-confidence. We may find that symbolic abstraction is a weakness as well as a strength. When it comes to problem solving, we might find that symbolic arithmetic is an antiquated tool. We might find that iconic arithmetic provides greater elegance, simplicity, groundedness and hopefully ease of learning.

In the rest of the volume, we will be seeing just how much of algebraic theory can be abandoned without loss, in pursuit of reconnecting mathematics with humanity. **Notation**, how we record mathematical ideas, is emphasized as both a problem and a solution.

In the text, we will often transcribe between two different representational (and conceptual) systems, between conventional strings and boundary math containers. I'll introduce a transcription symbol, ☞, to indicate when we are changing systems.

$$\textit{one formal system} \quad ☞ \quad \textit{another formal system}$$

We will restrict the use of ☞ to cases in which there is a specifically known **transcription map** between two systems, a map in which transcription does not introduce ambiguity. This makes ☞ a type of equal sign. We'll call it "*the finger*". The pointing finger is also a reminder that

Chapter 1

simple ≠ familiar

a *cognitive shift* is needed. Transcription is broader than mapping the same concept across two different notations. Underlying the profound difference between symbolic and iconic notations is a substantively different system of concepts about what arithmetic is and how it works. ☞ makes us aware that transcription from conventional expressions to boundary forms includes eliminating the conceptual infrastructure of sets, functions and logic.[45]

In *A New Kind of Science*, Stephan Wolfram valiantly reconstructs Science itself without calling upon calculus or algebra or arithmetic or logic.

> The presence of logic is in fact not essential to many overall properties of axiomatic systems.[46]

James algebra follows distantly in Wolfram's footsteps to construct a new kind of arithmetic. As you might expect, at first glance James patterns appear to be exotic (just like any math). The initial learning curve is not steep, but the new terrain may be quite unfamiliar.

To establish a common ground, in Chapter 2 we'll first visit a system that is familiar to us all, **unit-ensembles**. Unit-ensemble arithmetic describes how creatures like ///// behave. Then we'll tame the multiplicity of tally marks with the tools of depth-value in Chapter 3, and provide many sensory and dynamic depth-value representations of arithmetic in Chapter 4.

In Chapters 5 and 6, we'll become oriented to James algebra and the primary conceptual content of this volume. Chapters 7 through 12 address James algebra from a structural perspective. How can only containment express all of arithmetic? What makes a unit? How does structural change work? How does counting work? How do numbers work? In Chapters 13 and 14, we'll return to higher dimensions to provide several examples of sensory and experiential systems of James arithmetic.

Endnotes

1. opening quote: R. Feynman (1985) in K. Cole *Sympathetic Vibrations: Reflections on physics as a way of life.*

2. looking toward what Jaron Lanier calls postsymbolic communication: J. Lanier (2010) *You are Not a Gadget* p.190.

Lanier is motivated by the potential of virtual reality, a more visceral but still virtual interface with concept. "We'd then have the option of cutting out the "middleman" of symbols and directly creating shared experience. A fluid kind of concreteness might turn out to be more expressive than abstraction."

3. to assist with the tasks of arithmetic: Images on these pages are from the Computer History Museum, http://www.computerhistory.org/revolution/calculators/1

4. that gave numbers their current meaning: D. Schmandt-Besserat (1992) *How Writing Came About.*

5. arguments that they thought more secure: P. Davis (1997) Mathematics in an age of illiteracy. *SIAM News* **30**(9) 11/97 p.30.

6. solely to the kinds and orders of the symbols from which the expressions are constructed: R. Carnap (1937) *The Logical Syntax of Language* p.1.

7. the essential properties of the object to be represented: Hilbert's quotes are from M. Hallett (1994) Hilbert's Axiomatic Method and the Laws of Thought. In A. George (ed.) *Mathematics and Mind* p.184-185. Italics in the original.

8. what is generally known as the "axiomatic method": N. Bourbaki (1950) The architecture of mathematics. *American Mathematical Monthly* 57 p.222.

9. both the theory and practice of mathematics: J. Barwise & J. Etchemendy (1996) Visual information and valid reasoning. In G. Allwein & J. Barwise (eds.) *Logical Reasoning with Diagrams* p.3.

10. not necessary for a formal description of nor for an informal understanding of arithmetic: Throughout history mathematics has been associated with the work of God, although it is difficult to separate any of the activities of Western culture from attribution to a Christian diety. Philip Davis has a nice overview.

P. Davis (2004) A Brief Look at Mathematics and Theology. *The Humanistic Mathematics Network Journal Online* v27. Online 2/2017 at https://cs.nyu.edu/davise/personal/PJDBib.html

For incisive analysis, see B. Rotman (1993) *Ad Infinitum The Ghost in Turing's Machine: Taking God out of mathematics and putting the body back in.*

11. **mathematical thought exhibits organic diversity:** Ubiratan D'Ambrosio pioneered the idea that different cultures have different forms of mathematical thinking, contrasting the Western scientific method to more native styles of thought in South America. In A Histographical Proposal for Non-western Mathematics D'Ambrosio focuses on

> the social, political and cultural factors in the dynamics of the transfer and the production of scientific and mathematical knowledge in the colonies, as well as on the recognition of non-European forms of science and mathematics.

Quote in H. Selin (ed.) (2000) *The History of Non-western Mathematics* p.79-92.

A good analysis is in R. Vithal & O. Skovsmose (1997) The end of innocence: a critique of 'ethnomathematics'. *Educational Studies in Mathematics* 34 p.131-158.

12. **we live in many different worlds, rather than in just one world:** J. Goguen (1993) *On Notation.* Department of Computer Science and Engineering, University of California at San Diego.

13. **all reasoning depends directly or indirectly upon diagrams:** C.S. Peirce (1976) The new elements of mathematics. In C. Eisele (ed.) *Mathematical Philosophy* v4 p.314.

14. **within which are contained everything necessary for a correct inference:** D. Macbeth (2009) Meaning, use, and diagrams. *Ethics and Politics* xi(1) p.369-384.

15. **the parts make no sense without the whole:** G. Lakoff & R. Núñez (2000) *Where Mathematics Comes From: How the embodied mind brings mathematics into being* p.30-31.

16. **loading docks, suitcases, and kitchens:** In *Metaphors We Live By* (2003) G. Lakoff and M. Johnsen extend the CONTAINER metaphor to include houses, words, linguistic expressions, argumentation, forests, clouds, sports events, social groups, territoriality, time, activity, our visual field, our bodies, and our lives.

17. **its inside and its outside:** The Jordan curve theorem asserts that a closed loop on a flat surface does construct an inside and an outside. Its symbolic proof was notoriously difficult, presumably because the obvious properties of spatial forms are obscured by strings of symbols.

An engineer, a physicist, and a mathematician were faced with the problem of putting a herd of sheep inside a fence while using the least amount of fencing materials. The engineer rounded the sheep up into a tight group and then put a circular fence around them, declaring that the circle encloses the most area for the least fencing. The physicist imagined a fence with infinite radius, and then tightened it around the sheep, declaring that this fence is most efficient. The mathematician built a small fence around himself and declared: "I am on the outside".

18. **When we are inside a container, it feels like an environment:** Consider an automobile. When we stand outside it is a beautiful, or perhaps a utilitarian, object. When we drive, the automobile becomes an encompassing environment that is separate from the outside.

19. **Container schemas in the visual system:** Lakoff & Núñez, p.134.

20. **cannot reach the other side without crossing the boundary:** G. Spencer Brown (1969) *Laws of Form* p.1.

21. **Difference is not a quantity:** G. Bateson (1991) *A Sacred Unity* p.219.

22. **It is of the stuff of which minds are made:** Bateson, p.162.

23. **I only paint the difference between things:** attributed to Henri Matisse without a specific reference that I could locate.

24. **In a word, a difference is an idea:** G. Bateson (1972) *Steps to an Ecology of Mind* p.481.

Chapter 1

25. **but only differences and ideas:** Bateson, p.271.

26. **the logic of sets (sets contain their members), and so on:** Paraphrased from M. Johnson (1987) *The Body in the Mind*. In F. Varela, E. Thompson & E. Rosch (1991) *The Embodied Mind* p.177.

27. **It is part of embodied cognition:** Lakoff & Núñez, p.49.

28. **using the brain's cognitive mechanisms:** Lakoff & Núñez, p.3.

29. **its own autonomous laws of growth, its own dialectic:** I. Lakatos (1976) *Proofs and Refutations: The logic of mathematical discovery* p.146.

30. **the more extended is its area of applicability:** A. Einstein, in P. Schilpp (ed.) (1979) *Autobiographical Notes. A Centennial Edition* p.31. As quoted in D. Howard & J. Stachel (2000) *Einstein: The Formative Years, 1879-1909* p.1.

31. **the physics is inherent in mathematics:** Z. Artstein (2014) *Mathematics and the Real World* p.137-138.

32. **Beauty is the primary standard for scientific truth:** R. Augros & G. Stanciu (1984) *The New Story of Science* p.39.

33. **there is no permanent place in the world for ugly mathematics:** G. Hardy (1941) *A Mathematician's Apology* p.14.

34. **if they don't fit an elegant scheme, they are wrong:** M. Gell-Mann quoted in H. Judson (1980) *Search for Solutions* p.41.

35. **than to have them fit an experiment:** P. Dirac (1963) The evolution of the physicist's picture of nature. *Scientific American* **208**(5) p.45-53.

36. **to be taken as just that: a mathematical formalism:** R. Penrose (2004) *The Road to Reality* p.782.

37. **the theory should predict the results of measurements:** S. Hawking & R. Penrose (1996) *The Nature of Space and Time* p.121.

38. **the vast majority of student errors in math classes:** For example, see W. Bricken (1987) *Analyzing Errors in Elementary Mathematics*. Doctoral dissertation. Stanford University School of Education.

39. **links the abstract number to concrete notions of size, location and time:** S. Dehaene (2011) *The Number Sense: How the mind creates mathematics* p.246.

40. **geometry that already existed even in Babylonian times:** S. Wolfram (2002) *A New Kind of Science* p.792.

41. **not as justifications for mathematical rules:** A. Martínez (2006) *Negative Math* p.219.

42. **virtually nobody says that maybe it is the algebra that is defective:** Martínez, p.226.

43. **the associated subjectivities that permeate discourse — is crucial and nonnegotiable:** B. Rotman (1993) *Ad Infinitum* p.19.

44. **yet another mechanism of cultural imperialism:** The growth of big data has expanded this critique to include the impact of global quantification on our daily lives. "The bureaucratization of knowledge is above all an infinite excrescence of numbering." A. Badiou (2008) *Number and Numbers* §0.4

45. **eliminating the conceptual infrastructure of sets, functions and logic:** The root of almost all misinterpretations of boundary math is the attribution of concepts from conventional systems that do not exist within boundary systems.

46. **not essential to many overall properties of axiomatic systems:** Wolfram, p.1150.

Chapter 1

Chapter 2

Ensembles

Come, Mister tally man, tally me banana.[1]
— Harry Belafonte (1956)

What if arithmetic were simple? Really simple, so that addition and multiplication took no thought at all. The earliest arithmetic, at the pre-dawn of civilization at least 30,000 years ago, is called **tally arithmetic**. It is based on collecting together indistinguishable single markers such as pebbles, clay tokens, knots or strokes, with one tally mark corresponding to one tallied object. Philosopher of mathematics Brian Rotman:

Egyptian hieroglyphic script circa 3000 BCE

> Perhaps the most fundamental act of mathematical writing is the making of the Ur-marks or strokes in the form of patterns -- /, //, ///, ////, /////, and so on -- that correspond to the actual and imagined activity of counting.[2]

Tally arithmetic is really easy. Historian Georges Ifrah notes that tallying is "writing numbers in the simplest notation known."[3] To add, just put marks together. Putting together takes no cognitive effort, no thinking. Tally arithmetic uses the Additive Principle, that *a sum looks just like its parts*.

Chapter 2

2.1 Unit-ensembles

Unit-ensemble arithmetic is a formal arithmetic of strokes and tallies. The objects that we would associate with numbers are collections of units delineated by an outermost boundary. A **unit** is a mark, stroke, notch, pebble, shell, dot or other discrete singular indication of distinction. Here we'll use a dot, •, to represent one unit. Units may be replicated, providing a supply of *indistinguishable* units. Each unit is a **replica**. Replicas are intended to be identical in order to reduce the idea of counting to a foundation of one-to-one correspondence between marks and objects.

An **ensemble** is a collection of units within the same container. Ensembles

— are base-1. No grouping, no positional notation.

— use one-to-one correspondence. No counting.

— add by joining together.

— are the *definition* of the natural numbers.[4]

The magnitude of an ensemble is directly visible as the accumulation of tally marks. This magnitude is called the **cardinality**.

ensembles of 5

Our usual whole numbers are abbreviated names for ensembles. We prefer to say *five* rather than saying *one-one-one-one-one*.[5] Our common base-10 place-value notation is a grouping and naming convention for counting large collections of tallies. Numbers are **syntactic sugar** that make it easier to communicate magnitude, a convenience that does not interact with the rules and structure of arithmetic. The axioms of unit-ensemble arithmetic specify the behavior of both tallies and place-value numerals. The primary difference is that place-value requires linear textual thinking while unit-ensembles require iconic spatial thinking.

We're very familiar with unit-ensembles. Consider the American flag. The fifty identical stars form a unit-ensemble. They are placed symmetrically but the symmetry

Ensembles

Figure 2-1: *Whole numbers as ensembles*

has no intended meaning. Where stars are placed has changed over time. They are not in any particular groups and not in any particular ordering. The stars are in one-to-one correspondence with the States, but no particular star corresponds to a particular state. We can match stars with states and account for all of them without counting up how many stars or States there are. States and stars have the same cardinality.

The stars are added together by being put together on the blue field. The blue field is the container that makes the stars an ensemble. Similarly, the thirteen stripes are a unit-ensemble that represents the thirteen original colonies. No particular stripe corresponds to a particular colony. The six white stripes and the seven red stripes do form different ensembles but these colors have no meaning, they are artistic choices, presumably to help us to tell the stripes apart. The flag adds together the red and white ensembles, what we see is their *fusion*. A fusion is different than a sum because the thirteen stripes are no longer distinguished as separate once they are joined together as the flag. The stripes represent a unity, the Union of the original states.

Figure 2-1 shows whole numbers as unit-ensembles. The dashed lines represent the outermost containers that define each ensemble. Outermost containers can be either explicit or implicit. Just like set theory, a collection cannot be an ensemble without being contained within a boundary. The boundary, conceptual or physical, is what identifies the units within a collection as belonging together as a collection. Notice in Figure 2-1 that there is no zero. The absence of units means that there is nothing.

Chapter 2

Chinese rod digits circa 300 BCE

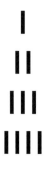

Units

In unit-ensembles *no unit is special*. The physical theory of relativity specifies that there is no special location within the universe, no perspective is privileged. The relativity of units denies that we can select a special unit, no unit is privileged. An alternative way to state this restriction is that no particular unit can be named. The identical units within an ensemble share absolutely no relative, metric or positional relationships. The only difference between replicated units is their independent presence. **Replicas** have separate existence but cannot be indexed, labeled or uniquely identified. Since units in an ensemble cannot be differentiated, they cannot participate in meaningful relations. Thus the ensemble is not defined by relationships between units in a group, but rather by the shared container.[6] This is an example of the boundary **Principle of Communality**.

Communality
If it is shared by all contents, it belongs to the context.

Our symbolic habits may mislead us about the nature of ensembles. The concept of the intersection of ensembles, for example, is not defined. If intersection were possible, then at least one unit could participate in two ensembles, thus imbuing that unit with a special, identifiable status. The properties of ensembles are isolated as properties of their container. A container itself has only one function, *containment*, and only one property, the cardinality of the units that it contains. This means that ensembles do not directly support ordinal numbers (first, second, third, etc).

We'll use a *tortoise-shell bracket*, ( ), as a value neutral outermost typographical container.[7] When two or more objects "share" the same space, the space itself does not support the relationship. Conventional notation invariably assumes an implicit operator in the space between objects. In a boundary form, that operator is the outermost container. When an outermost container is not recorded, the shell-bracket has been taken to be implicit.

Accumulation

In arithmetic units accumulate. Here's an algebraic assertion that units do not degrade in the presence of other units. Adding units makes an ensemble different.

$$(\bullet\ \bullet) \neq (\bullet)$$ **accumulation**

We could declare that replication of units is forbidden. Multiple units would condense into a singular identity. Adding units would not change the collection.

$$(\bullet\ \bullet) = (\bullet)$$ **idempotency**

This equation is an **idempotency** rule, and leads to a spatial Boolean algebra. Idempotency means that there is no additional effect when an object is recorded more than once, or when an operation is applied more than once. Idempotency is not unusual. Naming an object again is the same as naming it in the first place. Multiplying by zero again has no effect. Asserting the truth of a proposition again does not change its truth value. If we ignore social implication, sending an email twice is the same as sending it once. Including an object in a set more than once does not change the membership of the set. Idempotency is the reason that sets consist of unique members. Ensembles, in contrast, are not idempotent.

The constraint that the individual units within an ensemble cannot be distinguished from one another can be phrased in another, somewhat surprising, way: *we are not able to count units*. Counting requires assigning to each unit a unique identifier, that is, it requires making them distinguishable. Unit-ensemble arithmetic is sufficiently simple to exist prior to the idea of counting. The technique that replaces counting is **one-to-one correspondence**. Counting is, in fact, placing units in one-to-one correspondence with a sequence of unit increments. Human use of one-to-one correspondence precedes counting by tens of thousands of years.

The proposition that we do not need the structure of whole numbers for arithmetic sounds quite strange. However, when numbers are represented by unit-ensembles, we have the ensemble itself to stand in place of the unique integer it corresponds to. This approach is quite visual, an ensemble *illustrates* its cardinality. A natural concern would then be: how can we know the number of units in an ensemble without counting them? And of course, we cannot. The idea though is that we can separate the operations of arithmetic *calculation* from reading the magnitude of the result.

same

• • • • •
• • • • •

Containers with equal cardinality are equal. We do not need to count to verify equality.[8] The units in each container, instead, can be placed one beside the other with no leftovers. Mathematician Louis Kauffman:

different

• • • • •
• • • •

> It is not necessary to have a name for a number or to know how to count up to this number to apply the criterion of equality.[9]

comparison supports ordinal numbers

We still have a method of **comparison**: if the units of one ensemble do not match one-to-one with the units of another ensemble, then the two ensembles are not equal. We also know by what remains which ensemble is larger. One-to-one comparison is sufficient to construct an *ordering* of ensembles. Since units do not share relations and since all structure is determined by the container of the units, unit-ensemble arithmetic is a **boundary mathematics**.

Textual Notation

The spatial properties of ensembles lift the representation from the strict ordering of token-strings into the category of diagrams, drawings and other iconic forms. However, it is often convenient to have a typographic representation of iconic forms such as unit-ensembles, so we'll introduce a *textual notation*. A continuous line of dots can represent an ensemble in a line of text, although these are dot pictures, not words or expressions or numbers or sequences. Charles Peirce: "There are no better diagrammatic presentation of a number than a row of dots, all alike."[10] When we write a string of units, •••• for example, we are using an implicit boundary to delineate them. Explicitly, we could write (••••).

Although units do not support naming, ensembles do. We can name containers but not individual units because containers are differentiated by their contents. Later, within an interpretation as integers, we will label the diversity of types of ensembles with the numerals $\{1,2,3,\ldots\}$.[11] For now, capital letters $\{A,B,C,\ldots\}$ are used as names for particular ensembles, and small letters $\{a,b,c,\ldots\}$ are used to indicate the contents of particular ensembles. The capitalization distinction helps to avoid confusing the outside of a boundary with the inside.

Mayan digits circa 400 CE

•

••

•••

••••

2.2 Ensemble Arithmetic

Just like there is an arithmetic of numbers, there is an arithmetic of unit-ensembles. We can add and multiply and subtract and divide without having to count and without the idea of putting ensembles in order of size.

It's possible to build a formal theory of unit-ensemble arithmetic, similar to the theory of numbers built by Peano and Frege at the turn of the twentieth century. This chapter does not provide a complete formalism for unit-ensembles, however it does identify the relevant components.[12] Specifically, thus far:

— Containers have an inside, an outside, and a boundary.
— Units are indistinguishable.
— Units can be placed inside containers.
— A *whole number* is a collection of units in a container.
— Containers are equal only when their units are in one-to-one correspondence.

Chapter 2

Ensemble arithmetic at this point includes nothing else; no zero, no ordering, no inductive principle, no quantification, and in particular no logic, no sets and no functions. We can move directly to the definitions of addition and multiplication.

Addition

Unit-ensembles support the **Additive Principle**: the contents of ensembles contributing to a sum represent the sum. Unit-ensembles, and additive systems in general, can be either concrete or abstract. For example a pocket full of pennies is concrete, while making a mark on a piece of paper for each penny is abstract. Both copper pennies and marks on paper act additively.

1 ¾
means
1 + ¾

Residuals of additive notation are still present in today's arithmetic. Mixed numbers, for example, place whole numbers and fractions into the same space, implicitly embedding addition into their adjacency.

It is both historical and natural to think of addition as the act of putting things together. This perspective has recently been lost within the set theoretic operation of **union**, which combines the members of each set into a single set but requires members to be different. There is a simpler approach.

Figure 2-2 illustrates that unit-ensembles are added simply by *removing the boundary* between them. Addition can be achieved by (physical) deletion of separating boundaries, or it can be achieved by refocusing (cognitive) perspective. Unit-ensembles can be added together conceptually just by looking at them differently. Stop seeing different containers and start seeing a single container. There is no ordering when ensembles are added at the same time, and any number of ensembles add concurrently.

Figure 2-2: *Addition of unit-ensembles*

Fusion

The process of eliminating shared boundaries is fundamental to several boundary math systems.[13] **Mereology**, the study of parts and wholes, calls this process *fusion*. Fusion was the *general model of numeric addition* held by the mathematicians who created our current formal concepts of number over a hundred years ago, prior to the invention of set theory.

To add is to **fuse** containers. Mereological fusion is different than the conventional addition of numbers. In fusion, the original ensembles are lost. *Fusion creates one Whole*, one container. Nothing happens to the units, instead their context changes. Once the fusion occurs there is no information that remains to identify which units came from which component of the mereological addition. The former containers cease to exist. This is another way to state the Additive Principle: a whole is the fusion of its parts.

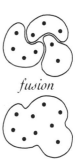

fusion

One major difference between mereological and set addition is that you cannot fuse nothing, an empty ensemble does not exist. In set theory, there is a difference between one object and the set of one object. Not so for fusion, there is only one type of thing, the ensemble. A single unit is also an ensemble of one. Units never stand alone, units are always bounded within an ensemble boundary. In our notation, one single unit can represent its own container.

Chapter 2

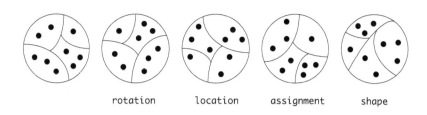

Figure 2-3: *Identical partitions*

Here is logician Gottlob Frege's definition. A **fusion**

Gottlob Frege
1848–1925

consists of objects; it is an aggregate, a collective unity, of them; if so it must vanish when these objects vanish...there can be no empty fusion.[14]

Each of the partitioned wholes in Figure 2-3 is identical. All fuse into the same object by removing the boundary partitions. The varieties pictured in Figure 2-3 illustrate visual distinctions that are not relevant to and are obliterated by fusion. Respectively, the rotation of a whole, the location of units within partitions, which partitions contain which units, and how a partition is shaped, each has no meaning when fused. The variety of appearance of these spatial representations is analogous to the variety of appearances of arbitrarily selected variable tokens within a string representation. From a functional perspective, we would say that fusion is a many-to-one mapping.

An operator that accepts any number of arguments is called **variary**. Variary operations do not have a specific **arity**, they do not require specific number of inputs. In particular, a whole is not formed by a sequence of binary fusions.

We are also leaving behind other features of modern function theory. Associativity conveys a temporal distinction, an artifact of operations that can accommodate only two arguments at one time, and therefore require sequential composition.[15] In a spatial model, this distinction is a syntactic implementation detail rather than a feature of the fusion operation.

Fusion can be read simply as *delete the separating boundary*. In our linearized dot notation,

$$(\bullet\bullet)(\bullet\bullet\bullet)(\bullet) \Rightarrow (\bullet\bullet\bullet\bullet\bullet\bullet)$$

Generally,

$$(a)(b)(c) \Rightarrow (a\ b\ c)$$ **fusion**

Fusion can also be expressed directly as a void-based transformation:

$$)(\Rightarrow void$$ **void-based fusion**

I've used the arrow rather than an equal sign to indicate that fusion has an asymmetric quality.

$$many \Rightarrow one$$

Another way to say this is that fusion is **lossy**. The partitions that are fused together are lost, there is no memory of which groups of units have been fused. The inverse of fusion is **partition**.

$$(a\ b\ c) \Rightarrow (a)(b)(c)$$ **partition**

Partition is one-to-many. A single fusion can be partitioned in many different ways. For example

$$(\bullet\bullet\bullet\bullet\bullet\bullet) \Rightarrow (\bullet\bullet)(\bullet\bullet\bullet\bullet)$$
$$(\bullet\bullet\bullet\bullet\bullet\bullet) \Rightarrow (\bullet\bullet)(\bullet\bullet\bullet)(\bullet)$$
$$(\bullet\bullet\bullet\bullet\bullet\bullet) \Rightarrow (\bullet)(\bullet)(\bullet)(\bullet)(\bullet)(\bullet)$$

We'll introduce a typographical abbreviation for clarity. A **bar** between units shows the potential to fuse by identifying different containers. Deleting the bar shows the result of the fusion. Here's Figure 2-2 in **fusion-bar** notation.

$$(\bullet\bullet\bullet|\bullet\bullet\bullet\bullet) \Rightarrow (\bullet\bullet\bullet\bullet\bullet\bullet\bullet)$$ *fusion-bar notation*

$$(a|b|c|d) \Rightarrow (a\ b\ c\ d)$$

Here's the general definition of finite fusion.

parallel fusion

$$(a|b|\ldots|z) \Rightarrow (a\ b\ \ldots z)$$

Having no concept of sequence, fusion is **flat**, it does not nest or iterate. Typographically,

$$((a|b)|(c|d)) = (a|b|c|d) \Rightarrow (a\ b\ c\ d)$$

That is to say, ensembles contain only units and not other ensembles. It may appear from a conventional perspective that temporal composition of fusions is not permitted, that we cannot take *steps* from a diverse beginning to a singular end. This is exactly what is intended. Temporal and operational steps are treated as occurring in parallel.

Similarly, fusion does not support a commutative rule. To add two sets, we usually begin with one set and we then add the other. This approach generates the need to identify a first set and second set, leading both to ordinal numbers and to the commutative rule:

$$A + B = B + A$$

It is not possible to state a commutative rule for ensembles. The equation

$$(\bullet\bullet|\bullet\bullet\bullet) = (\bullet\bullet\bullet|\bullet\bullet)$$

identifies a structural *identity*. Fusion is removing the shared boundary, there is no concept of position or ordering or quantity of units because it is the boundary that is being transformed, not the objects within the boundary.

$$(\bullet\bullet|\bullet\bullet\bullet) \Rightarrow (\bullet\bullet\bullet\bullet\bullet)$$

The illustration below shows that the commutative rule degenerates into a syntactic choice of rotation in a bounded space.

Ensembles

		0 ☞ void					
1	☞	(•)	–1	☞	<•>	=	(◊)
2		(••)	–2		<••>	=	(◊◊)
3		(•••)	–3		<•••>	=	(◊◊◊)
...			...				

Figure 2-4: *Integers to unit-ensembles*

Polarity

Symbolic notation uses the negative sign, –, to identify both inverse numbers such as –3 and the operation of finding a component of a sum by subtraction, as in 7 – 3. To be consistent with James algebra as presented in Chapter 10, we'll introduce a new type of boundary to serve as a negative sign and as the operation of changing the sign, or **polarity**, of an ensemble. The **angle bracket**, < >, identifies a "negative" ensemble. Figure 2-4 shows our current map from unit-ensembles to the integers.

attach "–"
–1   0   1

Written as <•>, the angle-bracket appears to be an operator on a positive unit, converting it from implicitly positive to explicitly negative. The angle bracket does serve as an outermost boundary, so that a shell-bracket is not needed.

<(•)> *is* <•>

<(a)> *is* < a >

We do have some subtle notational choices. Both shell- and angle-brackets are outer boundaries. We can represent a negative ensemble of magnitude A by <A>. Alternatively, we could explicitly construct a **negative unit** represented by a single token, ◊. The outer container is optional.

$$<•> \;=\; ◊ \;=\; (◊)$$

a single unit is its own container

The choice is whether or not to keep track of polarity on the outside or on the inside of the ensemble boundary. From the interior perspective, we must only keep clear the distinction that • and ◊ are marks, not ensembles.

Chapter 2

Reflection

Fusion applies only to the same type of boundary.

fusion
$$(a|b) \Rightarrow (a\ b)$$
$$<a|b> \Rightarrow <a\ b>$$

Two different types of boundary cannot fuse because they do not share a *common* boundary. By design, the Reflection axiom asserts that a unit and its reflection can be *deleted* from a shared container.

unit reflection
$$(\bullet\ <\bullet>) = (\ )$$

Unit Reflection is also expressed clearly using the negative unit token.

unit reflection
$$(\bullet\ \Diamond) = (\ )$$

In general, with the outer boundary implicit,

reflection
$$A\ <A> = \textit{void}$$

The purpose here is to show how negative numbers can be incorporated into unit-ensemble arithmetic. Before moving on to multiplication, a few observations are useful.

To subtract two arbitrary ensembles, it appears that matching the magnitudes of positive and negative forms is necessary. Ensemble subtraction however takes place via one-to-one correspondence between different types of *units*. Conventionally we might write something like this.

$$5 - 3 = (2 + 3) - 3 = 2 + (3 - 3) = 2 + 0 = 2$$

For ensembles this mechanism is unnecessary. There is no zero, but more importantly, the decomposition of units is already in place. Recall also that there is no ordering for units and no nesting, so Unit Reflection occurs throughout fused containers. The above numeric example incorporates three reflections, to look like this.

$$(\bullet\bullet\bullet\bullet\bullet\ |\ \Diamond\Diamond\Diamond) \Rightarrow (\bullet\bullet\bullet\bullet\bullet\ \ \Diamond\Diamond\Diamond) = (\bullet\bullet)$$

Second, the concept of a negative ensemble facilitates an immediate void-based mechanism for determining ensemble equality.

$$A = B \quad \Leftrightarrow \quad A <B> = \mathit{void} \qquad \textbf{void-based equality}$$

Void-based equality shifts the focus from comparison to existence. Should the form A <B> be meaningless, by ceasing to exist for example, then A = B.[16]

Finally, we'll need to consider reflected reflections, i.e.

$$<<\bullet>> = <\Diamond> = \bullet$$

We'll identify the structures associated with multiplying signed units a little later in the chapter. Strangely, it gets tricky.

2.3 Substitution

Substitution is the primary mechanism for computation and for proof in algebraic systems.[17] In general, Wolfram's *Mathematica* program demonstrates that practically all calculation, computation and symbolic demonstration can be achieved with pattern-matching and substitution over strings of symbols. Furthermore, when a mathematical system can be encoded in symbolic form, computation and demonstration in that system can be fully automated.

Substitution has always been difficult to show in mathematical notation because the form being substituted into changes when the substitution is enacted. Below, *double-struck tortoise shell brackets* are used as a compromise notation for the substitute operation.

$$\llbracket A\ C\ E \rrbracket \quad \text{☞} \quad \textit{substitute A for C in E}$$

Substitution is a ternary operation, involving three patterns that we have differentiated by their sequential position in the shell-bracket notation.

Chapter 2

⟦PUT FOR INTO⟧   ☞   *substitute* PUT *for* FOR *in* INTO

— The INTO-form is the structure that is being changed by the substitution.
— The PUT-form is the structure that is being substituted into the INTO-form, in the process changing the structure of the INTO-form.
— The FOR-form is the sub-structure within the INTO-form being replaced by the PUT-form.

Implicit within the substitution operation is an ability to locate or *match* the FOR structure within the INTO structure. This mechanism is far from trivial. The match might **fail**, in which case the substitution returns the INTO-form unchanged. The match might succeed more than once within the same INTO-form. This then requires several delicate implementation decisions that are considered in Volume II. In particular, it is critical whether substitution is partial (only some matches are changed) or total (all matches must be changed). When substitutions are nested or sequenced, additional implementation decisions are needed to decide how they might interact. At times during the course of a computation, for example, it is desirable to delay implementation of some substitutions until specific conditions are met.

before
E = (X Y Z)

during
E ⇒ ⟦Q Z E⟧

after
E = (X Y Q)

The form of substitution ⟦A C E⟧ is, like the √ symbol, an instruction rather than a result. With pattern-substitution, axioms depict a **before-and-after** picture. An equal sign is usually bidirectional, it doesn't matter which side of the equation we consider as *before*. In contrast, substitution shows us a specified *before*, the INTO form E. The result is the entire expression ⟦A C E⟧. To maintain consistency, all C in E must be replaced by A. Substitution must be total, considering all content forms as well as their contents.

Substitution includes several structural identities that define its behavior in any algebra. Figure 2-5 shows these structural and value-preserving symmetries. The

STRUCTURAL INVARIANCE

$(\!($ A E E $)\!)$ = A   **global substitution**

$(\!($ A A E $)\!)$ = E   **self substitution**

REPLACEMENT

$(\!($ A C E $)\!)$ = E   *given* A = C   **value maintenance**

SUBSTITUTION EQUALITY

$(\!($ A C E $)\!)$ = $(\!($ B C E $)\!)$ *given* A = B   **put-equality**

$(\!($ A C E $)\!)$ = $(\!($ A D E $)\!)$ *given* C = D   **for-equality**

$(\!($ A C E $)\!)$ = $(\!($ A C F $)\!)$ *given* E = F   **into-equality**

VOID-SUBSTITUTION

$(\!($ *void* C E $)\!)$ ⇒ *delete* C *from* E   **delete**

$(\!($ A *void* E $)\!)$ ⇒ *construct* A *within* E   **construct**

Figure 2-5: *Substitution*

behavior of substitution is often bundled with the definition of equality as "substitute equals for equals". This however is only one of the invariants of substitution, the one that maintains value but not structure.

Figure 2-5 begins with two structurally invariant substitutions, followed by the common replacement of equals by equals. Next there are three equality maintenance cases when PUT-, FOR- and INTO-forms are equal. There are also two special void-substitution cases. **Deletion** is substituting *void* for a given form. **Construction** is substituting a given form for *void*. Void-substitution maintains the value of E only when C and A are both void-equivalent. Finally, substitution distributes across equality.

$(\!($ A C E=F $)\!)$ ⇔ $(\!($ A C E $)\!)$ = $(\!($ A C F $)\!)$ **distribution of equality over substitution**

Chapter 2

Multiplication

Multiplication of unit-ensembles is achieved by replacing every unit in one ensemble by an *ensemble*. The **Multiplicative Principle** is that every unit from one ensemble interacts with every unit from another ensemble. The interaction can be pairing (using ordered pairs), touching (using grids lines), or in the case of ensembles, replacing each unit with an ensemble, using substitution followed by fusion.

(a b c) × (d e)
is
(a d) (a e) (b d)
(b e) (c d) (c e)

$$A \times E \quad \Rrightarrow \quad (\!(A \bullet E)\!)$$

To multiply, substitute A for every unit in E. For example, *substitute 4 for each unit in 3* would be written as

$$(\!(\,(\bullet\bullet\bullet\bullet) \quad \bullet \quad (\bullet\bullet\bullet)\,)\!)$$

or more symbolically as $(\!(4 \ 1 \ 3)\!)$.

The process of substitution can be illustrated by a sequence of structural changes in the typographic representation of multiplication-as-substitution.

initial form	$(\!(\,(\bullet\bullet\bullet\bullet) \quad \bullet \quad (\bullet\bullet\bullet)\,)\!)$
unit partition	$(\!(\,(\bullet\bullet\bullet\bullet) \quad \bullet \quad (\bullet\|\bullet\|\bullet)\,)\!)$
substitute	$(\bullet\bullet\bullet\bullet\|\bullet\bullet\bullet\bullet\|\bullet\bullet\bullet\bullet)$
fuse	$(\bullet\bullet\bullet\bullet\bullet\bullet\bullet\bullet\bullet\bullet\bullet\bullet)$

Figure 2-6 shows the same substitution process using a diagrammatic notation.

The closest that positional notation comes to multiplication-as-substitution would be to decompose a number into units, and then distribute over units.

$$4 \times 3 = 4 \times (1 + 1 + 1) = 4 + 4 + 4 = 12$$

Positional notation, naturally, does not permit substitution within the sequence of digit positions. Instead, we distribute the multiplication over units and add the result.

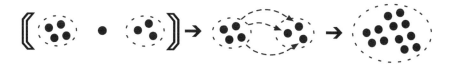

Figure 2-6: *Multiplication of unit-ensembles*, 4×3

Unit-ensembles also have a specific distribution theorem, distributing fusion over substitution.

⟦A C E|F⟧ = ⟦A C E⟧ | ⟦A C F⟧   **distribution of fusion over substitution**

Distribution asserts that ensembles can be fused first and then substituted into, or they can be substituted into first and then fused. For example:

⟦(•••) • (•|••)⟧ = ⟦(•••) • (•)⟧ | ⟦(•••) • (••)⟧

3 × (1 + 2) = (3 × 1) + (3 × 2)

Commutativity

The unique property that differentiates substitution as a generic process from substitution that models multiplication is *substitution symmetry* across PUT- and INTO-forms. This is conventionally called the **commutativity of multiplication**.

⟦PUT • INTO⟧ = ⟦INTO • PUT⟧   **commutativity**

With PUT/INTO symmetry, the semantics of substitution provides all the remaining structural axioms that define the behavior of multiplication. From the perspective of substitution, this constraint is quite unusual. We would usually expect to find the PUT-form within the INTO-form, however here we also expect to find the INTO-form within the PUT-form. With unit-ensembles, this constraint is easily met, since all ensembles are composed of identical units.

Chapter 2

Although the PUT-form and the INTO-form are symmetric with regard to the result of a substitution, the *process* of substitution does distinguish differential effort. For example, consider the simple case of 2 x 3 vs 3 x 2:

$$2 \times 3 \; ☞ \; ((•••) \; • \; (••)) = (•••|•••) \Rightarrow (••••••)$$
$$3 \times 2 \; ☞ \; ((••) \; • \; (•••)) = (••|••|••) \Rightarrow (••••••)$$

The computational effort is in the first substitution step, not in the second fusion step, since fusion is blind to magnitude. Effort is required to locate matches prior to being able to substitute.

Division

The same substitution structure used as a model of multiplication provides a model for unit-ensemble division.

$$E \div C \; ☞ \; (• \; C \; E)$$

To divide an ensemble E into portions of magnitude C, substitute • for every C in E. Here's an example of an exact division of units.

$$6 \div 3 \; ☞ \; (• \quad (•••) \quad (••••••))$$

partition $\quad (• \quad (•••) \quad (•••|•••))$

substitute $\quad (•|•)$

fuse $\quad (• \; •)$

Substitution-as-division imposes the same PUT/INTO symmetry as does substitution-as-multiplication.

symmetry $\qquad (• \; \text{FOR INTO}) = (\text{INTO FOR} \; •)$

The difference between multiplication and division, from the perspective of substitution, is in the location of the unit, not in the symmetry of substitutions. Multiplication has a unit FOR-form while division has a unit PUT-form.

There is a problem. Consider this example after applying commutativity of substitution.

$$6 \div 3 \ \text{☞} \ (\!(\ \bullet \ (\bullet\bullet\bullet) \ (\bullet\bullet\bullet\bullet\bullet\bullet) \ )\!)$$
$$(\!((\bullet\bullet\bullet\bullet\bullet\bullet) \ (\bullet\bullet\bullet) \ \bullet \ )\!) \quad \text{commute}$$

Since there is no (•••) in •, there is no match to trigger a substitution. Here we see the introduction of a new imaginary form, the fraction. Yes, from the perspective of a unit whole, *fractions are imaginary*. We have no notation yet for parts of a whole. Even the option of calling upon commutativity is eliminated by the form of a unit fraction.

$$1 \div C \ \text{☞} \ (\!(\ \bullet \ C \ \bullet \ )\!) \quad \textbf{unit fraction}$$

This situation is analogous to conventional notation, for which 1/3, for example, is also written as an operation yet to be enacted.

$$1/3 \ \text{☞} \ (\!(\ \bullet \ (\bullet\bullet\bullet) \ \bullet \ )\!)$$

To complete the structural symmetry, here is the form of squaring.

$$A \times A \ \text{☞} \ (\!(A \ \bullet \ A)\!)$$

These substitution patterns expose the communality of negative (C^{-1}) and positive (A^2) exponents as variation of the location of a unit-ensemble within the substitution operation.

More generally, substitution integrates both multiplication and division into the same operation.

$$A \times E \div C \ \text{☞} \ (\!(A \ C \ E)\!) \quad \textbf{integrated substitution}$$

It is relatively easy to extend the substitution model of division to include remainders.

Kauffman has emphasized that a substitution is one step of a recursive process.[18] Within unit-ensembles with substitution he has exposed embedded concepts of fixed-point functions, fractal geometry, self-referential systems and paradox.

Chapter 2

natural substitution	failed substitution
$(\bullet\ \bullet\ \bullet) = \bullet$	$(\bullet\ \diamond\ \bullet) = \bullet/\diamond$ *to-be-decided*
$(\bullet\ \diamond\ \diamond) = \bullet$ *commutes*	$(\diamond\ \diamond\ \bullet) = \bullet$
$(\diamond\ \bullet\ \bullet) = \diamond$ *commutes*	$(\bullet\ \bullet\ \diamond) = \diamond$
$(\diamond\ \diamond\ \diamond) = \diamond$	$(\diamond\ \bullet\ \diamond) = \diamond/\bullet$ *to-be-decided*

Figure 2-7: *Options for substitution of signed units*

Polar Multiplication

We are now ready to tackle briefly the once controversial rules of multiplication of signed, or polar, numbers.[19] *Why* does multiplying a negative times a negative result in a positive? Initially it is sufficient to consider substitution of only the signed units, • and ◊. Figure 2-7 shows the eight possible substitution patterns for signed units. Four patterns are direct substitutions without ambiguity. Four others are substitutions that fail to find a match and therefore do not change the INTO form. Two of the failures commute into natural substitutions while maintaining a consistent result.

Remaining are two patterns identified as *to-be-decided*. Both fail to find a match. The design choice of what to do in these two circumstances was controversial for hundreds of years. Which defining characteristics of multiplication are necessary to support our conception of what multiplication is? As Martínez observes, one possible design decision is to abandon commutativity of multiplication, and develop the rules of signs based on other principles.[20]

Here we will take as valid only substitutions that make sense within the match-and-substitute paradigm. Since the defining property of multiplication-as-substitution is commutativity, we'll take as fundamental the idea that

Ensembles

substitution is commutative. Commutativity is mute for the substitution forms marked *to-be-decided*, since their PUT-forms are the same as their INTO-forms.

We're left with four possible options for interpreting the two *to-be-decided* substitutions.

Option 1: ⟦● ◊ ●⟧ ⇒ ●   ⟦◊ ● ◊⟧ ⇒ ◊
Option 2: ⟦● ◊ ●⟧ ⇒ ◊   ⟦◊ ● ◊⟧ ⇒ ●
Option 3: ⟦● ◊ ●⟧ ⇒ ●   ⟦◊ ● ◊⟧ ⇒ ●
Option 4: ⟦● ◊ ●⟧ ⇒ ◊   ⟦◊ ● ◊⟧ ⇒ ◊

I'll caution that these assignments are choices, and each option can lead to reasonable systems of arithmetic with unfamiliar but not contradictory behavior.

Only *Option 1* makes sense intuitively as a substitution, both substitutions fail to find a match. *Option 1* is to assert that failure to match results in no change of the INTO-form. Interpreted, however, *Option 1* says that plus divided by minus is plus and that minus times minus is minus.

Option 2 is counter-intuitive from the perspective of failure-to-match and would require a new type of substitution rule to override the failed match. In terms of our interpretation as multiplication, it is *Option 2* that aligns with the conventional rule of signed multiplication by Leonard Euler, that "minus times minus is plus".[21] *Options 3* and *4* abandon symmetry and are compromises that satisfy neither consistency nor the rules of substitution. I'll pass on considering these two options.

One step toward resolution of the choice between *Options 1* and *2* is to return to substitution of ensembles (rather than signed units). ◊ is an abbreviation for <●>. We can limit application of Reflection to ensembles, while abandoning the concept of a discrete negative unit.

Chapter 2

Consider the substitution form ⟦◊ • ◊⟧. We'll replace ◊ by <•> and carry out the direct substitution.

$$\llbracket \text{<•>} \bullet \text{<•>} \rrbracket \Rightarrow \text{<<•>>}$$

Should we assert that this direct substitution is valid, then the two options provide these two choices.

Option 1:   <<•>> = <•>

Option 2:   <<•>> = •

Option 1 shows that a double angle bracket is not different than a single angle-bracket. Angle-brackets are idempotent.

Option 2 shows that double angle-brackets cancel. We can also demonstrate that *Option 2* is a consequence of the Reflection axiom.

```
initial form            <<•>>
     reflect            <<•>> <•>  •
     reflect                        •
```

Another technique to resolve the question of multiplication of negative numbers is to call upon the associativity relation. We could assert that multiplication-as-substitution is **associative**. When a substitution operation is embedded within another substitution operation, the choice of which substitution to do first is quite flexible. The assertion of commutativity imposes an extreme freedom upon the associativity of substitutions. Within our notation,

associativity   ⟦⟦A C E⟧ D F⟧ = ⟦A ⟦C E D⟧ F⟧ = ⟦A C ⟦E D F⟧⟧

If we assume *Option 1*, then we have this consequence.

```
                • = ⟦•        •      • ⟧
     substitute     ⟦• ⟦• ◊ •⟧ • ⟧
      associate     ⟦• • ⟦◊ • •⟧⟧
     substitute     ⟦• •      ◊      ⟧ ⇒ ◊
```

In contrast, *Option 2* does not generate a contradiction. The bottom line is that *Option 2*, "minus times minus is plus", maintains commutativity and associativity, at the cost of abandoning the model of multiplication-as-substitution for the specific case of $(\!(\diamond \bullet \diamond)\!)$.

Option 1 degrades the associativity of multiplication, as Martínez demonstrates using conventional expressions.[22] When we adopt *Option 1*, the symbolic model of multiplication (still based on substitution and on the same distributive rules) requires both anticommutativity and non-associativity to remain consistent.

anticommutativity
A x B = − B x A

A surprising result of *Option 1* is that − x − = − unifies imaginary numbers with real numbers, since $\sqrt{-1} = -1$.

$$\begin{aligned} i &= \sqrt{(-1)} \\ &= \sqrt{(-1 \times -1)} \\ &= \sqrt{(-1)} \times \sqrt{(-1)} \\ &= (\sqrt{-1})^2 \\ &= -1 \end{aligned}$$

This little excursion does not resolve the question of which rule of signed multiplication is preferable. It does depend upon context and objective. The design decision comes down to a choice of structural preferences. We have defined addition and multiplication by fusion and substitution respectively in order to insist upon a physical basis for physical arithmetic. In Martínez' analysis of the historical and structural foundations of arithmetic, he concludes:

> But if we desire a symbolic system that serves not only as a method of calculation but also as a mode of representation, then we should strive to develop an algebra in which every operation can be understood in physical terms.[23]

The conclusion we can draw for this short exploration is that from any perspective, the concept of a **negative number** is unavoidably inconsistent with the principle of anchoring arithmetic to physical reality.

2.4 Comparison

Figure 2-8 shows a comparison of conventional group theoretic axioms for arithmetic to those of unit-ensembles. There is structure and function hidden within each notation, so the comparison is not intended to demonstrate conceptual simplicity. Rather the comparison shows the understanding needed in order to conduct the operations of arithmetic in each of these abstract theories. Although we have been associating structural manipulation with numeric multiplication, the conventional axioms of multiplication and addition have not been used in the development of unit-ensembles. Figure 2-8 also shows a hybrid system in which the axioms of group theory are expressed in the language of unit ensembles. With a little boundary thinking, a more intuitive, physically natural axiom structure for integers and the operations of arithmetic is available. Volume II compares the boundary mathematics approach to the currently popular formal definitions of numbers by Frege, Peano and other twentieth century philosophers of arithmetic.

2.5 Remarks

In the course of building formal mathematics on a basis of set theory, the naturalness of addition by putting things together was lost. In schools, at least until college, the structure of set theory is not taught explicitly, leaving a collection of seemingly meaningless constraints like associativity and commutativity as the "rules of arithmetic". These are not the rules of arithmetic, they are more properly the rules of a set theoretic interpretation of how numbers work when expressed as strings of symbols.

Unit-ensemble arithmetic provides a simple example of how addition and multiplication do not necessarily require complex algorithms and symbol manipulation. To add, put ensembles together. To multiply, put one ensemble inside another.

Ensembles

Unit-Ensemble

	addition	*multiplication*
fuse	(A|B) = (A B)	
commute		⟦A • E⟧ = ⟦E • A⟧
reflect	A | <A> = *void*	

Group Theory

	addition	*multiplication*
commute	x + y = y + x	x*y = y*x
associate	(x+y) + z = x + (y+z)	(x*y)*z = x*(y*z)
identity	x + 0 = x	1*x = x
inverse	x + −x = 0	x*(1/x) = 1
distribute	x * (y+z) = (x*y) + (x*z)	

Group Theory as Fuse-and-Substitute

	addition	*multiplication*
commute		⟦A • E⟧ = ⟦E • A⟧
associate		⟦A • ⟦B • E⟧⟧ = ⟦⟦A • B⟧ • E⟧
identity	E = E	⟦• • E⟧ = E
inverse	A | <A> = *void*	⟦A • ⟦• A •⟧⟧ = •
distribute	⟦A • E|F⟧ = ⟦A • E⟧ | ⟦A • F⟧	

Figure 2-8: *Axiom systems for arithmetic*

A central motivation for this chapter is to introduce a substantive portion of the structure of James algebra. Although the upcoming James model of addition is slightly different than fusion, and the James model of multiplication is substantively different than substitution, the boundary structure of both systems is remarkably similar because addition and multiplication in each system is simply an interpretation of specific forms of containment. The shared structure includes:

— the Additive Principle
— addition by putting together in the same container
— the round-bracket as the positive unit
— the angle-bracket as a generalized negation
— void-based equality
— Reflection, A <A> = *void*

We will make only one substantive modification in the shift from unit-ensembles to James algebra, and that is to imbue *units* with an interior. James units are hollow. Interiors allow us to convert a unit object into an operator just by putting forms inside its boundary.

Unit-ensemble arithmetic is supported by several web-pages at the iconicmath.com website.

Webpage: iconicmath.com/arithmetic/units/
The **Unit-ensembles** webpage uploaded in 2012 is a preliminary draft of the structural rules in this chapter.

Webpage: iconicmath.com/algebra/containers/
The **Container Algebra** webpage includes a discussion of unit-ensemble arithmetic. The page provides

details of substitution of equals, fusion (called *merging* on the webpage), and multiplication-as-substitution. There's a suggestion to use positive and negative signs as *indicators of change* rather than for value management. The behavior of fractions as substitution forms is discussed in depth.

We have one more step to take prior to the introduction of James algebra, and that is to address the issue of reading the magnitude of large numbers of unit tallies.

Endnotes

1. **opening quote:** Lyric from a traditional Jamaican folk song, circa 1900, made popular by Harry Belafonte in his 1956 signature song, *Day-O (The Banana Boat Song)*.

2. **the actual and imagined activity of counting:** B. Rotman (2000) *Mathematics as Sign: writing imagining counting* p.ix.

3. **writing numbers in the simplest notation known:** G. Ifrah (2000) *The Universal History of Numbers* p.xix.

Ifrah shows that in every culture, the numbers 1, 2, and 3 began as one, two, and three strokes. The origin of these simple numerals is tally marks. The shapes of our numbers, at least the first few, are derivative of these tallies.

4. **is the definition of whole numbers:** D. Schmandt-Besserat (1987) Oneness, Twoness, Threeness. *The Sciences* 27 p.44-48.

5. **say five rather than to say one-one-one-one-one:** In *Through the Looking Glass*, (L. Carroll, 1871) Alice and the Red Queen have this conversation:

> "And you do Addition?" the White Queen asked. "What's one and one and one and one and one and one and one and one and one and one?" "I don't know," said Alice. "I lost count." "She can't do Addition," the Red Queen interrupted. "Can you do Subtraction? Take nine from eight." "Nine from eight I can't, you know," Alice replied very readily: "but--" "She can't do Subtraction," said the White Queen.

6. **but rather by a shared container:** Of course, the word *shared* takes liberty with the absence of properties between units. Indistinguishable units cannot participate in a shares relation. Only the container has the cardinality property.

7. **tortoise-shell bracket, ( ), as a value-neutral outermost typographical container:** The more common delimiters — the parenthesis ( ), the square-bracket [ ], and the angle bracket < > — are used as part of the James algebra notation, so the relatively unique tortoise-shell is used here to avoid confusion. The shell-bracket ( ) is used solely as a value-neutral, void-equivalent outermost container throughout the volume. Notice that tortoise shell brackets can be applied anywhere, not necessarily to both

sides of an equation. It makes no sense to apply a tortoise shell bracket to a single object since the object provides its own outermost bracket explicitly. Figure 5-2 lists the types of brackets used in the text.

8. We do not need to count to verify equality: The idea that equal contents means equal cardinality has a long philosophical history. **Hume's Principle** asserts (roughly) that two collections have equal numerosity (are equinumerous) only when they can be put into one-to-one correspondence. Briefly,

Hume's Principle
Numeric equality is one-to-one correspondence.

When Frege was developing our contemporary definition of what a number is, he took the controversial position that numbers could be derived completely from pure logic. Hume's Principle is the linchpin. Unfortunately the principle does not tell us what a number is, instead it tells us what equality is. And unfortunately for logicians, the principle is not a statement within first-order logic. Incidentally, Hume did not create this principle, Frege did. It's called Hume's Principle cause Frege introduced it with a quote from Hume. A good resource for the mathematical and logical controversies surrounding Hume's Principle is G. Boolos (1998) *Logic, Logic, and Logic*.

The generally accepted definition of number is due to Russell: *a number is a collection of all sets equinumerous with it.* In his article Definition of Number, in J. Newman (1956) *The World of Mathematics* p.537-543, Russell asserts "The number of a class is the class of all those classes that are similar to it." We'll explore how we came to this in Volume II. Oh, and on a lighter note, Frege invented a now abandoned word for equal numerosity. It translates from German as *identinumerate* or *tautarithmic*.

9. count up to this number to apply the criterion of equality: L. Kauffman What is a Number? Online 6/2016 at http://homepages.math.uic.edu/~kauffman/NUM.html

10. than a row of dots, all alike: C. Peirce (MS. 179, 1872) in V. Garnica *Changes and Chances: an initial study of Peirce's pragmatism and mathematical writings as they relate to education and the teaching and learning of mathematics* p.22. Online 1/2017 at http://socialsciences.exeter.ac.uk/education/research/centres/stem/publications/pmej/pome19/index.htm

Ensembles

11. **types of ensembles with the numerals {1,2,3,...}:** The curly brace { } is used exclusively in its conventional sense as a notation for sets.

12. **we will identify the relevant components:** When integer arithmetic was being formalized at the beginning of the twentieth century, Hilbert and Dedekind were clearly thinking about tally arithmetic. Hilbert thought it too obvious to formalize. The formal structure of tally arithmetic is known as Presburger arithmetic. These stories are told in Volume II.

13. **fundamental to several boundary math systems:** Several alternative systems are included in Chapter 14.

14. **there can be no empty fusion:** G. Frege quoted in M. Potter (2004) *Set Theory and its Philosophy* p.23.

15. **accommodate only two arguments at one time, and therefore require temporal composition:** An abstract mathematical expression such as a+(b+c) can be interpreted as a *completed* addition but for a single step. Identifying two additions does not necessarily imply two simplification steps. It does imply two structural distinctions, otherwise the parentheses are not needed. Without an ability to count, we cannot identify *two* steps; both must occur outside of time and outside of structure. Structural changes are usually identified as steps counted using ordinal numbers (first and second) which do not require the inductive whole number sequence.

16. **Should the form of A <B> be meaningless, then A = B:** We can call upon Reflection to demonstrate void-based equality as a theorem.

$$A = B$$
$$<B> A = B <B>$$
$$<B> A = void$$

The structure and properties of equality, including the ubiquitous permission to do the same thing to both sides of an equation, are discussed in Volume II.

17. **the primary mechanism for computation and for proof in algebraic systems:** There's a more extensive discussion in Volume II.

18. **substitution is one step of a recursive process:** L. Kauffman (1986) *Formal Arithmetic*. Department of Mathematics, Statistics and Computer Science, University of Illinois at Chicago.

Chapter 2

19. **controversial rules of multiplication of signed, or polar, numbers:** This historical controversy is detailed in Volume III.

20. **develop the rules of signs based on other principles:** A. Martínez (2006) *Negative Math* Chapter 7.

21. **"minus times minus is plus":** L. Euler *et al* (1822) *Elements of Algebra* p.8. The exact phrase is *"− minus* multiplied by *− minus* gives *+ plus."*

22. **as Martínez demonstrates using conventional expressions:** Martínez, Chapter 6.

23. **every operation can be understood in physical terms:** Martínez, p.207.

Chapter 3

Depth

[The positional number system] is also the ultimate perfection of numerical notation...no further improvement of numerical notation is necessary, or even possible.[1]
— *Georges Ifrah (2000)*

We have ignored until now a fundamental problem: how do we represent large numbers in an additive system? If 346 units and 274 units are put together, how can we efficiently count up the result? Unit-ensemble arithmetic has a distinct drawback: it's very inconvenient to determine how many marks there are in a large ensemble. It is certainly convenient to fuse many large piles by pushing them all together, but that does not help to determine *how many?*. The time-honored solution is to **group units** into bundles. Grouping tally marks was added to historical systems of calculation as perhaps the first elaboration to tallying at least 20,000 years ago.[2] The Babylonians used groups of sixty. The Romans used groups of five and ten. English money before 1970 was grouped in twelves and twenties.

Our conventional number system uses positional notation to simplify counting units by grouping them into bundles of ten. The **uniform base** of ten permits us to construct bundles of bundles of ten to get bundles of one hundred,

Chapter 3

5283 *is*
$5000 = 5 \times 10^3$
$200 = 2 \times 10^2$
$80 = 8 \times 10^1$
$3 = 3 \times 10^0$

and so on. We stack units and tens and hundreds from right to left, the reverse of normal Western reading patterns. To know how big a number is, we need first to scan to the right to count or estimate the number of digits, and then backtrack to the left for the magnitude, hiding this complexity behind years of training. Chapter 11 considers the diversity of representations of number in detail.

The cost of base-10 positional notation is **cognitive and computational load**. There are one hundred digit addition facts to memorize. If you understand the structure of the addition table, this reduces to less than twenty, however understanding structure also requires cognitive training and effort. Almost half of the addition facts require *carrying* (that is, the sum is greater than 9). The addition and multiplication algorithms taught in grade school add their own, somewhat onerous, overhead.[3]

What is taught as addition is an amalgam of number facts, group theory, number theory, notational structure, bundle management, and pragmatic algorithms for knitting place-value numerals together.[4] In contrast, unit-ensembles separate theory, structure and algorithm by using the generic tools of pattern-matching to enact transformations. We have also reverted to the original theory of numbers as **additive ensembles of units**. To isolate the structural mechanism of grouping units, we'll introduce a new type of boundary that supports manipulation of *groups*. The parenthesis ( ), called **parens**, identifies groups and their base. But a word of caution, the parenthesis-bracket will have a different definition when we get to James algebra.[5]

Depth-value is independent of both unit-ensembles and James forms. It is a boundary *extension* to either system that adds the capability to assemble units into bundles of a specific magnitude. Depth-value notation uses depth of nesting rather than left-to-right position to keep track of grouping, elegantly integrating boundary notation for large numbers with boundary operations on those numbers.[6] This simple yet substantive change illustrates the disadvantages of having a notation for numbers (place-value, circa 800 CE) that was not specifically designed to be compatible with today's common operations of calculation (symbolic counting, addition and multiplication), all of which are recent additions.

We'll finish this chapter with **parens arithmetic**, a new formal system based upon depth-value notation. Parens arithmetic maintains

- addition as fusion
- multiplication-as-substitution
- subtraction as fusion augmented by reflection
- division as pattern substitution.

These techniques were first published in 1995 by Louis H. Kauffman, a Professor of Mathematics at the University of Illinois at Chicago, and a pioneer in boundary mathematics and in topology. The linear textual version of depth-value should properly be called **Kauffman numbers**.[7]

3.1 Order of Magnitude

The **order of magnitude** of a digit within a conventional number identifies how many times that digit is multiplied by ten. Instead of increasing the power of a digit by moving to the next position on the left (thousands, hundreds, tens, units), depth-value increases power by depth of nesting. Crossing a boundary outward changes the order of magnitude by one power-of-ten, more generally by one power of an arbitrary base. 5283 in positional notation is (((5)2)8)3 in depth-value notation. The depth-value form of 5003 is (((5)))3. The absence of a digit between two boundaries can be taken to mean zero. There is no need for an explicit 0 token to fill in missing magnitudes when no digit is present.

0	☞	
1		1
9		9
10	(1)	
12	(1)	2
20	(2)	
100	((1))	
253	((2)5)3	
804	((8))4	
4000	(((4)))	

Chapter 3

a place-value integer	5283				
positional notation	5		2	8	3
polynomial form	5×10^3	+	2×10^2	+ 8×10^1	+ 3×10^0
factored form	$(((((5 \times 10) + 2) \times 10) + 8) \times 10) + 3$				
embed base-10	(((5	) + 2	) + 8	) + 3	
+ *is containment*	(((5	) 2	) 8	) 3	
a depth-value integer	(((5)2)8)3				

Figure 3-1: *Place-value to depth-value*

Do not be concerned by the extra burden of brackets. Positional notation has the same burden. By convention we do not write the meanings of the left-to-right positions within the number. Instead we spend years as children learning to see them implicitly. With familiarity, the brackets too can be implicit. For example, the leading open parentheses can easily be omitted to yield, for example,

5)2)8)3   ☞   5283

Figure 3-1 shows an example of the conversion from place-value to depth-value notation. Place-value is an abbreviation for numbers rendered as polynomial expressions, while depth-value is an abbreviation for numbers rendered as factored expressions. For convenience, we'll consider both ensemble and depth-value forms to be *implicitly bounded* by a value-neutral container.[8]

Both place-value and depth-value accommodate any whole number other than 1 as the base. (The tally system is base-1.) The depth-value form of a place-value number can be constructed by repeated factoring of the the base and then deleting both the multiplication by the base and the addition signs. The base is absorbed into the interpretation of the parens boundary. The order

$$\bullet\cdot\cdot_N\cdot\cdot\bullet = (\bullet) \qquad \textbf{group}$$

$$(a)(b) = (a\ b) \qquad \textbf{merge}$$

Figure 3-2: *Depth-value transformation patterns*

of magnitude is absorbed into depth of nesting while the addition sign is absorbed into the featureless space within each bracket.

Place-value notation has the power of the base *implicitly* lined up with textual columns while depth-value has the power of the base *explicitly* recorded as the depth of nesting. This allows computation to proceed at all levels at the same time, converting sequential dependencies into parallel freedoms.

3.2 Standardization

The great advantage of depth-value notation is that it is explicitly defined by two simple transformation rules, shown in Figure 3-2. To **group** is to construct bundles of units, while to **merge** is to join these bundles. These two rules lay the basis for the integration of the representation of magnitude with the operations of arithmetic, creating *parens arithmetic*.

The numeric concept of a *base* is easier to understand in the context of group and merge, since these two depth-value operations separate the maintenance of groups from the operations of arithmetic. It's easier to change bases since the grouping rule makes how bases work explicit. Both group and merge can be construed as physical actions. You might recognize merge as ensemble fusion.

Chapter 3

Group

The **group** transformation converts a collection of units into one unit nested one level deeper. For base-10, ten units are converted into one nested unit.

base-10 *group*
$$\bullet\bullet\bullet\bullet\bullet\bullet\bullet\bullet\bullet\bullet = (\bullet) \;\;\reflectbox{☞}\;\; 10$$

In the case of base-2, the group operation is

base-2 *group*
$$\bullet\bullet = (\bullet) \;\;\reflectbox{☞}\;\; 2$$

In general, for any base-N,

unit group
$$\bullet.._N..\bullet = (\bullet) \;\;\reflectbox{☞}\;\; N$$

Group makes explicit any implicitly assigned base for a parens form, similar to what we do when we gather, say, twelve eggs and put them into a container to make a dozen. The implicit base is defined by the egg-carton. More generally, any pattern can be grouped.

group
$$A.._N..A = (A) \;\;\reflectbox{☞}\;\; A \times N$$

As is apparent, the group transformation is an alternative way to express multiplication. In Chapter 9, group will become Replication, a fundamental component of the theory of counting. The pattern that converts tallies into numbers is a special case of group for which $A = 1$.

$$\bullet.._N..\bullet = (\bullet) \;\;\reflectbox{☞}\;\; 1 \times N$$

Merge

We'll need a second conversion rule that combines *groups* together to consolidate forms that share the same level of nesting. Instead of units, merge manages the aggregation of boundaries that have the same base.

merge
$$(a)(b) = (a\;b)$$

More elegantly,

$$)(\;\;=\;\; \textit{void}$$

Boundary merging is *spatial* so it is not easily shown in linear text. It is the physical process we use to collect groups rather than units.

Merge has the same definition as fuse: *delete common boundaries*. The difference between the two is their generality. Fusion joins unit-ensembles. Merge applies to any boundaries that share a common grouping rule, regardless of contents. Fusion emphasizes that the joined contents form a whole, while merging places no expectation on the newly formed contents. More technically, fusion is unidirectional, while merging is symmetric and bidirectional.

Like fusion, merge also applies to any number of boundaries in the same context concurrently. More generally then,

$$(a)\ldots(z) = (a\ldots z) \qquad \textbf{parallel merge}$$

Canonical Form

Depth-value is a syntactic standardization technique for enhancing the readability of numbers. It does not directly impact the objects, operations or concepts of conventional arithmetic. That is, unless we consider group and merge to *be* the operations of arithmetic, with group as multiplication and merge as addition.

Canonical form identifies a *unique* structure for every expression with the same value. Each sum of whole numbers, for example, has a single whole number that is canonical for that sum. The purpose of standardization, whether place-value or depth-value, is to identify canonical form.

> *Common arithmetic is how we find the canonical form of an expression within place-value notation.*

Each boundary number has two canonical forms. The **canonical unit form** is a unit-ensemble consisting only of units. It has the advantage of having no internal structure

canonical unit five
1+1+1+1+1

varieties of five
1+1+3
2+1+2
2+3
(2x3)−1
√25

canonical symbol five
5

Chapter 3

```
1 ☞   •
2 ☞   •• = (•)
3 ☞   ••• = (•)•
4 ☞   •••• = (•)(•)        = (••)    = ((•))
5 ☞   ••••• = (•)(•)•      = (••)•   = ((•))•
6 ☞   •••••• = (•)(•)(•)   = (•••)   = ((•)•)
7 ☞   ••••••• = (•)(•)(•)• = (•••)•  = ((•)•)•
8 ☞   •••••••• = (•)(•)(•)(•) = (••••) = ((•)(•))
                                      = ((••)) = (((•)))
```

Figure 3-3: *Depth-value standardization*, base-2

unit form
••••••••

depth-value form
(((•)))

and the disadvantage of requiring effort to determine how many?. The **depth-value form** is the result of all possible merge and group operations. It has the minimal number of units required to express a given magnitude, and the maximal depth of nesting. The depth-value form closely resembles our universal place-value notation but has a significantly different functional structure.

Figure 3-3 shows the application of group and merge to the depth standardization of tallies. The grouping rule is shown in base-2.

A canonical form is *efficient but not necessary* for arithmetic operations to proceed. Elementary school arithmetic is dominated by the standardization process of writing numbers in what is taught to be their most convenient form. How arithmetic operations actually work is generally ignored. Standardization itself does not necessarily construct the most convenient form of a number, since *convenience is determined by use, not by rule*.[9] Standardization is also partial, since expressions such as fractions, 2/3 for example, maintain an operator within the number itself. Analogously, 2x3 is a number, not standardized to the canonical 6 but just as legitimate as 2/3.

Depth

Reading

Canonical forms are easiest to read. A depth-value number can be read in two different ways. We can follow the conventions of place-value but reading from deepest to shallowest. Or we can read constructively. Begin at the innermost content and head outward. Multiply the accumulated contents by the embedded base each time a boundary is crossed. Add any numbers in the context crossed into. Here's a base-10 example.

1 *to* 16 *in* base-2
depth-value

•
(•)
(•)•
((•))
((•))•
((•)•)
((•)•)•
(((•)))
(((•)))•
(((•))•)
(((•))•)•
(((•)•))
(((•)•))•
(((•)•)•)
(((•)•)•)•
((((•))))

((4)3)2	☞	432
4		4
(4)	*times ten is*	40
(4)3	*plus three is*	43
((4)3)	*times ten is*	430
((4)3)2	*plus two is*	432

There is no ordering principle within boundary forms. Each of these typographic boundary forms is the same.

((4)3)2 = 2((4)3) = (3(4))2   ☞   432

It is better though to visualize a two-dimensional display in which forms roam anywhere, limited only by their surrounding boundaries.

To read a standardized base-2 parens form, begin with the unit in the deepest nesting as 1. Double as you cross outwards over each boundary. Add 1 if there is a unit in the context crossed into. Forms nested inside a boundary are thus doubled in value by their boundary.[10] Here's a base-2 example, the number 22.

22 ☞ ((((•))•)•)		
•		1
(•)	x2 *is*	2
((•))	x2 *is*	4
((•))•	+1 *is*	5
(((•))•)	x2 *is*	10
(((•))•)•	+1 *is*	11
((((•))•)•)	x2 *is*	22

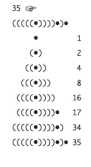

35 ☞
(((((•))))•)•

•	1
(•)	2
((•))	4
(((•)))	8
((((•))))	16
((((•))))•	17
(((((•))))•)	34
(((((•))))•)•	35

3.3 Parens Arithmetic

Parens arithmetic explicitly separates the operations of arithmetic from the standardization of how a number is written or displayed. The primary change is to think of conventional arithmetic operations like addition or multiplication as extremely simple: one parallel put *together* for addition, one parallel put *inside* for multiplication. Separate from the operations of arithmetic is what we might normally call *computation*. Parens computation constructs a **canonical form** by alternating applications of parallel group and merge. Computation is implemented by one-to-one pattern-matching and substitution.

to add:
put together outside

to multiply:
put together inside

Animation

In text, each parens form is a static, linearized diagram. Sequences of changes to the parens form can be visualized as animations by showing the *after* form of a transformation directly underneath the *before* form. The textual animation shows time and change moving down the page, line by line. Each line identifies a pattern of containers and units. Figure 3-4 shows how each transformation is represented by before and after changes in parens forms. The figure also shows the inverse transformations unmerge and ungroup. Although create, the inverse of cancel, is included for symmetry, standardization provides no motivation to create polarized pairs from *void*.

before ●●
    group
after (●)

before (●)(●)
    merge
after (●●)

Both merge and cancel are deletion rules that are particularly easy to read. Structure is deleted as steps progress down the page. In each case the structure being deleted is present directly above and absent below.

Addition

Addition is putting forms together. Both unit tallies and grouped tallies add by being placed together in the same (implicit) container. The alternation of parallel merge

Depth

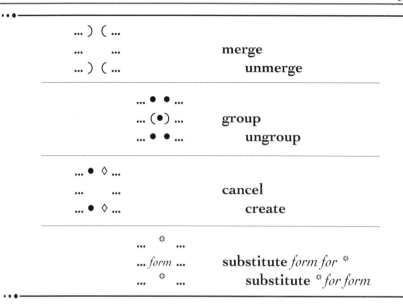

Figure 3-4: *Animation of parens transformations*

followed by parallel group generates the canonical form of the sum. Here is a base-2 example.

7 + 5 ☞ ((●)●)● ●((●))		put 7, 5
((●)●)(●)((●))		group
((●)● ● (●))		merge x2
((●)(●) (●))		group
((● ● ●))		merge x2
(((●) ●)) ☞ 12		group

The addition of 7 + 5 is completed on the first line. The value of the first line is 12, but the form that results after putting 7 and 5 together is not optimal. It corresponds to the conventional expression 6 + 1 + 1 + 4. The numbers 6 and 4 are pre-simplified by the canonical forms of 7 and 5.

Figure 3-5 shows a base-10 addition using both unit-ensembles and their digit abbreviations. The digit abbreviations necessarily call upon number facts to add digit symbols. This depth-value standardization process is displayed in a linear, string-based format. As a

Chapter 3

```
    432 + 281 ☞

••  (•••  (••••))((••)  •••••••••)•     2 (3 (4))((2) 8)   1
••  (•••  (••••      ••)  •••••••••)•   2 (3 (4    2) 8)   1
((•••• ••)  •••  •••••••••) •• •          ( (4 2) 3  8) 2 1  ←
((•••• ••)  (•)          •) •• •          ( (4 2)(1) 1) 2 1
((•••• ••    •)          •) •• •          ( (4 2  1) 1) 2 1
                                          ( (   7   ) 1) 3
            ☞   713
```

Figure 3-5: *Parens addition*, base-10: 432+281

consequence, the step marked by the arrow in Figure 3-5 is a linear *reordering* to help reading and not part of the parallel merge process. In contrast, Figure 3-6 shows the same addition in a planar spatial notation, the fluid circle dialect explored in Chapter 4.

Figure 3-7 shows another example of base-2 parens addition, with many more parallel operations. The figure adds three numbers concurrently, illustrating that putting parens numbers together (i.e. adding them) is insensitive to how many numbers there are. We are essentially doing parallel column addition. At the top of the figure the boundaries of the three numbers to be added are aligned by depth. All boundaries are then merged concurrently, resulting in 18 different merge operations if they are counted one-at-a-time. The rest of the figure shows alternating group and merge standardization.

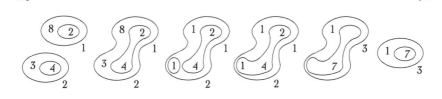

Figure 3-6: *Fluid circle addition*, base-10-digit: 432+281

Depth

```
                                                                    align
(((((((((((        •)•)      ))•) )    •)) )•)          1618
(((((((((          •)) ) )         ))•) )    •           261
(((((((((((((•)) ) )•    )) )•)•    )) ) )•             4401

(((((((((((((•))•)•)• •))•)•)• •))•)•)• •              merge x 18
(((((((((((((•))•)•)(•)))•)•)(•)))•)•)(•)              group x 3
(((((((((((((•))•)•   •)))•)•   •)))•)•    •)          merge x 3
(((((((((((((•))•)(•  )))•)(•   ))))•)(•  ))           group x 3
(((((((((((((•))•  •  )))•  •   ))))•  •  ))           merge x 3
(((((((((((((•))(•  ) ))))(• )  ))))(•  )  ))          group x 3
(((((((((((((•)   •   )  ))) •  )  )))   •  )  ))      merge x 3

(((((((((((((•)•))))•))))•)))                          6280
```

Figure 3-7: *Parens parallel addition*, base-2: 1618+261+4401

Subtraction

Depth-value can be extended to include negative numbers by creating a new type of unit. We introduced the negative unit ◊ in Chapter 2, and it behaves the same as a positive unit.

```
 2   ☞     •• = (•)
 1   ☞     •
 0   ☞     void
-1   ☞     ◊
-2   ☞     ◊◊ = (◊)
-3   ☞     ◊◊◊ = (◊)◊
-4   ☞     ◊◊◊◊ = (◊)(◊) = (◊◊) = ((◊))
-N   ☞     ◊..N..◊
```

Unit Reflection is the only new rule necessary to implement addition and multiplication of negative numbers.[11]

$$• \; ◊ \; = \; void$$ **unit reflection**

In general, subtraction is addition of a form after changing the polarity of its units. Here is a simple example of depth-value subtraction in base-2. An empty grouping, ( ), is void-equivalent.

put	5 – 7 = 5 + –7 ☞	• ((•)) ((◊) ◊) ◊
merge		• ((• ◊) ◊) ◊
reflect		(( ) ◊)
empty group		(◊) ☞ –2

Figure 3-8 provides an example of subtraction of depth-value forms in base-10. The significant difference is that negative units carry their polarity at the unit level rather than at the magnitude level. For example,

–281 ☞ ((◊◊) ◊◊◊◊◊◊◊) ◊

To change the polarity of a number, change the unit type. The example in Figure 3-8 includes *a new type of number* that is not available in place-value expressions, the **mixed-polarity** form ((2)–5)1. To standardize mixed-polarity numbers, one unit of opposite polarity in the next deeper level is unmerged and ungrouped, the depth-value equivalent of *borrowing*. This is the same process taught in schools, but made explicit and decoupled from specific place-values. Thus, sequences of borrowing which place a mental burden on a student turn into independent manipulations. The following base-10 example shows three instances of ungroup.

	1234 – 987 ☞	
put	4 (3 (2 (1))) ((–9)–8)–7	
merge	4 (3 (2 (1)) –9)–8)–7	
reorder	(((1) 2 –9) 3 –8) 4 –7	
cancel	(((1) –7) –5) –3	
ungroup	((10 –7) –5) –3	
cancel	((3) –5) –3	
unmerge	((2)(1) –5) –3	
ungroup	((2) 10 –5) –3	
cancel	((2) 5) –3	
unmerge	((2) 4) (1) –3	
ungroup	((2) 4) 10 –3	
cancel	((2) 4) 7	

☞ 247

Depth

```
    432 - 281  ☞
●● (●●● (●●●●))((◊◊) ◊◊◊◊◊◊◊◊) ◊        2 (3 (4)) ((-2)-8)-1
●● (●●● (●●●●    ◊◊) ◊◊◊◊◊◊◊◊) ◊        2 (3 (4    -2)-8)-1
((●●●● ◊◊) ●●● ◊◊◊◊◊◊◊◊) ●● ◊           (  (4 -2) 3 -8) 2 -1
((●  ●  )              ◊◊◊◊◊) ●         (  (2   )    -5) 1
((●)(●   )             ◊◊◊◊◊) ●         (  (1)(1)    -5) 1
((●) ●●●●●●●●●●        ◊◊◊◊◊) ●         (  (1) 10    -5) 1
((●) ●●●●●                   ) ●         (  (1)  5      ) 1
              ☞  151
```

Figure 3-8: *Parens subtraction*, base-10: 432–281

The larger example in Figure 3-9 shows the concurrent addition of both positive and negative numbers. The standardization process is insensitive to the polarity of numbers. The six integers in Figure 3-9 are added by 29 concurrent merges followed by 6 parallel cancels.[12]

```
                                                              align
(((((((         ●)  ●)      ●))    ●)   )  ●)                 234
(((((                    ◊ ))   ◊ )  ◊)   )  ◊                - 45
(((((((● )  )● )         ))  ●   )   )   )●                   329
(((((((    ◊ )   )  ◊  ))  ◊   )   ) ◊ )◊                    -171
(((((((  ◊  )   )◊      ))       )◊  )   )                   -164
(((                            ●   )   )●  )                   10

(((((((● )◊◊●)● ●)◊ ◊◊●))●◊●●◊●)◊ ◊)●◊●)◊●◊     merge x 29
(((((((● )◊  )● ●)◊  ◊   ))●    )◊ ◊)   ●)◊     cancel x 6
(((((((● )◊  )(●))(◊)   ))●    )(◊))    ●)◊     group x 3
(((((((● )◊    ●)   ◊)  ))●       ◊))   ●)◊     merge x 3
(((((((● )        )  ◊) ))          ))   ●)◊     cancel x 2
((((((( ●  ●      )  ◊) ))          ))   ●)◊     ungroup
(((((((●)(●       )  ◊) ))          )))(●)◊     unmerge x 2
(((((((●)●           ●  ◊) ))       )))●  ●◊     ungroup x 2
(((((((●)●                )  ))     )))●         cancel x 2
((((((((●)●))))))●                                193
```

Figure 3-9: *Parens parallel addition*, base-2: 234–45+329–171–164+10

Chapter 3

Multiplication

Multiplication-as-substitution combines seamlessly with the group and merge process. To multiply, substitute a parens form for each unit in another parens form. Neither form needs to be canonical. Non-canonical forms are just less efficient, requiring more group and merge operations to standardize.

Here is the base-2 multiplication 2 x 3. Multiplication is implemented by substituting the form of 3 for each unit in the form of 2.

	2 x 3 ☞ *Put* (•)• *into* (•)	
form of 2	(*)	
substitute	((•)•)	☞ 6

The form of 2 incorporates only one unit, so one substitution takes place. The result is directly the form of 6, no merging or grouping is needed.

Here's a simple example that requires standardization after substitution, 3 x 3, again in base-2.

	3 x 3 ☞ *Put* (•)• *into* (•)•	
form of 3	(*) *	
substitute	((•)•) (•)•	☞ 6 + 2 + 1
merge	((•)• •)•	
group	((•)(•))•	
merge	((• •))•	
group	(((•)))•	☞ 8 + 1 = 9

In this example, once the substitution has taken place on line two, multiplication has been achieved. The form on line two has the value of the product, in this case 9. In the four steps that follow, two pairs of merge and group implement standardization of the product.

The next two examples show 5 x 7 and 7 x 5.

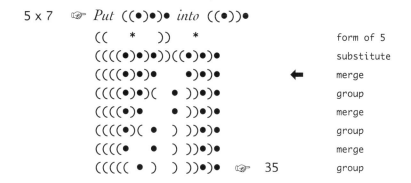

Substitution of ((•)•)• into ((•))• completes the multiplication step. Simplification then reduces both commutative variants to the same canonical form.

```
7 x 5    ☞  Put ((•))• into ((•)•)•
            ((    *    )    *    )    *                form of 7
            ((((•))  •)((•))  •)((•))•                 substitute
            ((((•))  •   (•))  •   (•))•               merge
            ((((•))(•)   •  )(•)   •  )•               reorder
            ((((•)   •)   •   •)   •   )•        ←     merge
```

The group and merge process for both of the commutative variants converges after a few steps. At the rows marked by the arrow, the partially simplified form of 7 x 5 is identical to the form of 5 x 7. Unlike conventional notation, we can see clearly the difference in computational effort imposed by the order of multiplication. The put of 5 into 7 requires one additional merge to standardize.

3.4 Remarks

In parens arithmetic, addition and multiplication are essentially trivial, they involve only the effort of placing forms in particular locations. The effort of standardization is to improve readability, just like we might assemble coins into groups of ten in order to simplify counting a large pile of pennies. The entire algorithmic structure of parens arithmetic spans three pattern rules (group, merge, reflect) and one operation (put). All rules and operation are achieved by substitution. Chapter 4 demonstrates these rules in their spatial dynamic forms.

> *Iconic Mathematics*
> *math that looks like what it means*

The iconicmath.com website includes extensive descriptions, examples, calculators, videos and animated tutorials on depth-value notation and parens arithmetic.

Introduction: iconicmath.com/arithmetic/depthvalue/

The **Depth-value** webpage describes the forms and algorithms of depth-value notation.

Animation: iconicmath.com/arithmetic/containers/

The **Container Numbers** page is the heart of the depth-value section of the site. It includes pointers to narrated and short animated tutorials and examples in several visual dynamic notations that are described in Chapter 4. The page includes about 30 short tutorial animations of the rules of depth-value.

Linear notation: iconicmath.com/arithmetic/parens/

The **Parens Notation** page contains dozens of examples of parens integer arithmetic, with a focus on the interactive dynamics of group, merge and reflect. The examples are arranged in a tutorial sequence. Each parens example corresponds to an example from the container arithmetic page iconicmath.com/arithmetic/containers/

Iconic calculator: iconicmath.com/calculators/iconiccalculator/

The **Iconic Calculator** page is a portal to the three implemented modes of the iconic calculator. The calculator uses depth-value to show reduction steps for many examples of base-2, base-10-unit, and base-10-digit arithmetic.

Depth-value, along with parens arithmetic, demonstrates that there is an alternative to conventional arithmetic that is both efficient and intuitive. James arithmetic, however, goes further, showing us new ways to think about the rational and irrational numbers.

Next, we'll look at the dynamic forms of parens arithmetic. Then Chapter 5 summarizes the structural (but not the conceptual) content of James algebra. Chapter 6 provides an advanced organizer that describes what to expect when encountering James algebra. Do not expect that this initial exposure will rest easily on the eyes at first glance. A different way of thinking about numbers is needed to make the boundary notation familiar. We are heading into a territory in which elementary mathematics is about putting things into containers. Nothing more.

Endnotes

1. **opening quote:** G. Ifrah (2000) *The Universal History of Numbers* p.346.

2. **the first elaboration of tallying at least 20,000 years ago:** Several separate examples of grouped notches carved on a bone have been discovered in archaeological digs dated circa 30,000 BCE. G. Ifrah (2000) *The Universal History of Numbers* p.62.

3. **add their own, somewhat onerous, overhead:** A classic grade school challenge is to ask why students need to learn the long division algorithm. In earlier decades as a student I had to learn to compute square roots by hand. As these arcane practices fell away to hand-held computing devices, the question still remains: how much paper-and-pencil computing do students need to know? The brave among educators concede that long multiplication may no longer be necessary, but most draw the line at column addition. In this volume, I'm suggesting that even *counting* may be too much. More importantly, it is not which algorithms to teach that is important, it is rather whether or not our current algorithms for arithmetic operations are learnable at all by more than one person in three. Educator Marilyn Burns popularized these issues in (1998) *Math: Facing an American phobia*.

4. notational structure, and pragmatic algorithms for knitting place-value numerals together: The addition and multiplication algorithms taught in grade school are not necessarily modern, nor are they tailored for ease of learning nor for cognitive efficiency. They are best thought of as rather random evolutionary variations that have accumulated by chance and by fashion. Different countries use different methods for determining sums and products, as have different historical ages. I believe that how people added and multiplied in the 18th century could be superior to how we are taught today. For example, multiplication was done by successive doubling. Folks mastered the two-times table and that was sufficient. Two hundred years ago, the techniques of arithmetic had to be tailored to cognitive needs because folks had little skill and even less experience with arithmetic. Today's algorithms have been distorted by the rush to symbolic approaches a century ago, embracing an abstract ideal while ignoring evolutionary precedence, native human skills, and neurological organization. Keith Devlin (*The Math Gene*, 2000, p.60.):

> There must be something about the multiplication tables that makes them all but unlearnable. Such a widespread problem with multiplication surely indicates some feature of the human brain that requires investigation, not criticism.

5. a different definition when we get to James algebra: Beginning in Chapter 5, we'll redefine the parenthesis to be the basic numeric unit in the James system, calling it the *round-bracket*. The parenthesis is the only typographical delimiter that changes meaning across two different systems.

6. integrating boundary notation for large numbers with boundary operations on those numbers: The idea of nesting to convey the exponent of a number was present in Roman numerals. What we identify now as letters (I, V, X, L, C, M) were not *letters* to the Romans, but rather particular symbols, probably derived from marking tally sticks. I is not the letter I, it is a notch in a stick. V and X are each types of tally notches that are easy to distinguish when the tally stick is split in half. M, one thousand, evolved from CIƆ, which itself began as a circle drawn round a vertical stroke. The Romans used CƆ to stand in place of *parentheses*. Thus depth-value notation was in use over two thousand years ago. The Roman numeral for 10,000 is CCIƆƆ, while 100,000 is CCCIƆƆƆ. F. Cajori (1928) *A History of Mathematical Notations* §51.

Roman tally stick

7. **depth-value should properly be called Kauffman numbers:** Some of Professor Kauffman's many innovations in boundary mathematics are described in Chapter 14.

8. **implicitly bounded by a value-neutral container:** To avoid any confusion between explicit and implicit outermost containers, the *tortoise-shell bracket*, ( ), was used in Chapter 2 as a value-neutral container for ensembles. It was not used in this chapter since the outermost neutral boundary is taken to be implicit.

9. **convenience is determined by use, not by rule:** Standardizing a numeric expression also loses information about how that number came to be. For example, our checkout bill at the supermarket includes both the standardized total cost, and a list of the costs of each item. Both are useful for different reasons. The total is handy for paying the bill, the itemized list is necessary for verifying correct charges. The disconnect between memorizing standardization rules in elementary school and the legitimate use of numbers to describe events is one of the central reasons that math confuses students. No other curriculum subject is built upon a disconnection from reality.

10. **Forms nested inside a boundary are thus doubled in value by the boundary:** The doubling and halving method was a standard way to do multiplication and division throughout history. It is found on the Rhind papyrus from the Second Intermediate Egyptian Period, 1650 BCE. It is also know as Russian Peasant Multiplication and is apparently still in use.

11. **to implement addition and multiplication of negative numbers:** Reflection in this case is stated without the structure of ensembles since the grouping operation provides that structure. Fusion is also absent, replaced by an implicit global container.

12. **added by 29 concurrent merges followed by 6 parallel cancels:** The count of these merges is an artifact of notation. There are 29 separate merge operations in the example in Figure 3-9, *if merge is read as changing fragments of text*. There are 8 depth merges if we count sequential steps rather than individual textual instances of the rule. If we read the aligned forms as merging vertically, then all 29 of the parallel merges occur at one time. There is effectively *one* merge operation.

Chapter 3

As well, the figure artificially separates the application of different rules as an aid to reading, but they too can occur in parallel. After the three `merge` operations on line 4, the `cancel` and `ungroup` operations on the following two lines could have occurred concurrently. Two different patterns applied at the same time to two different locations in a string is still one parallel transformation in a spatial form.

Chapter 4

Dynamics

...essentially every piece of graphical notation that anyone's ever tried to invent for anything seems to have had the same problem of being hard to understand.[1]
— *Stephen Wolfram (2007)*

Now we'll look at some **spatial dialects** for depth-value. A primary feature of depth-value is that addition and multiplication require very little effort, specifically one parallel put operation. Alternating applications of group and merge can then optimize the result to a form in which nesting is maximal and the number of units is minimal.

Depth-value incorporates the Additive Principle and multiplication-as-substitution. Extending depth-value with a negative unit provides negative numbers, while substitution easily extends to division. It's straight-forward to read depth-value as conventional arithmetic in a somewhat different notation. However, depth-value is substantively different in two ways. First, it covers the structure and transformations of arithmetic using three simple pattern axioms; one simple operation, put; and one generic implementation mechanism, match-and-substitute. Second, in this chapter we'll see that depth-value arithmetic can be expressed in higher dimensional, interactive forms.

Chapter 4

CANONICAL FORM

$\bullet\cdot\cdot_N\cdot\cdot\bullet = (\bullet)$	**unit group**
(a)(b) = (a b)	**merge**
● ◊ = *void*	**unit reflect**

INTERPRETATION

A B	☞ A + B	**put**
〚A C E〛	☞ (A x E)/C	**substitute**

Figure 4-1: *Depth-value patterns*

4.1 Variety

Figure 4-1 shows the axioms and operations of parens arithmetic, the entire formal structure needed for the arithmetic of rational numbers.

In this chapter, we'll be exploring substitution pattern rules as dynamic transformation of spatial forms. Two of the *pattern rules* identify void-equivalent structure, an instruction to remove, eliminate, or delete form. The *two operations* are specifically actions: putting things together and replacing one thing by another. Of course, the flat page of a text cannot show manipulation of physical blocks, for example, except as an image that might serve as an instruction to act. Each parens transformation takes on a different look and feel for the four different spatial dialects that follow.

— one-dimensional textual parens (Chapter 3)
— two-dimensional circular enclosures
— three-dimensional blocks
— dimension-free networks

Most of the time I'll show grouping in base-2, since that creates less cluttered images. I'm going to finesse the problem of *describing* the dynamics of these dialects by providing sequences of stop-action animation frames without much narration, and by referring to the website iconicmath.com that has videos of the animations.

4.2 Circle Arithmetic

Circle arithmetic is a visualization of container arithmetic with two-dimensional oval enclosures providing boundaries. It is simply parens arithmetic expressed on a plane with the fractured textual delimiters returned to complete enclosures.

Animations: iconicmath.com/arithmetic/containers/

> The **Arithmetic Containers** webpage includes dozens of video animations of 2D container arithmetic arranged in a tutorial sequence.

Along the sidebar, frames of the animation of 7 + 5 are shown in stop-action.[2] In Figure 4-2 on the following page, there are frames from the circle animation of 5 x 7 on the left, and 35 ÷ 5 on the right. The animations track the same group and merge steps as the parens examples in Chapter 3.

Fluid Boundaries

Figures 4-3 and 4-4 (on pages 92 and 93 respectively) show the steps of 319 x 548 expressed as base-10-digits. In order to use digits in a boundary form, we need to call upon the digit multiplication facts, which are shown in hybrid form within *Frame 3* of Figure 4-3. The use of digit abbreviations rather than unit-ensembles imposes a cost: we need to know the meaning of the arbitrary digit symbols in combination with one another, in other words, we need the digit multiplication table.

Frame 1 of Figure 4-3 shows the fluid boundary forms of 319 and 548. In *Frame 2*, the form of 319 is enlarged in order to accommodate the form of 548 at each depth.

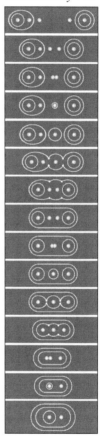

15 frames from the circle animation of 7+5

Chapter 4

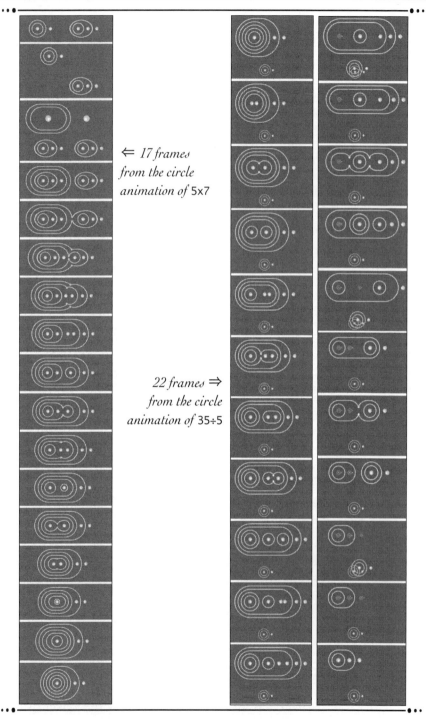

Figure 4-2: *Circle dialect:* 5×7 *and* 35÷5

Dynamics

The specific digits of 319 at each depth multiply the entire 548 form. *Frame 2* is hybrid, with the multiplication sign showing. This particular frame is included to show pattern substitution better, and does not occur during boundary multiplication. Instead, substitution proceeds in parallel directly to *Frame 3* in which the digits of 319 each descend into every level of 548.

Frame 3 shows the two step parallel percolation of digits into their appropriate depth. The *global form* of 548 pervades each level of nesting in 319. The *local digit* of 319 at that level then pervades each level of nesting in 548. This implements the Multiplicative Principle, that each part of one form multiplies (touches) each part of another form. Not showing in between *Frame 3* and *Frame 4* is **unit multiplication** in which numeral labels are replaced by unit-ensembles. The *unit level* provides a third pervasion in which every *unit* in each pervading digit touches every unit of the digit being multiplied. Touching is implemented by unit-ensemble substitution. For example,

fluid merge

```
3 x 4   ☞   ( 4   •   3 )
            (••••  •  •••)              form of 3 x 4
            (••••  •  •|•|•)            partition
            ••••|••••|••••              substitute
            ••••••••••••                fuse
            (  •  ) ••                  group
                    ☞  (1) 2            hybrid
```

Frame 4 shows the result of the digit multiplication as either a single digit or in the form (digit)digit. After *Frame 4*, we see alternating parallel merge and group transformations that terminate when no further pattern-rules are triggered.[3] Fluid boundaries merge by flowing together, and then disappearing when they touch. After each merge, the digits in each separate context are replaced by their addition fact, again either as a single digit or in the form (digit)digit.

Chapter 4

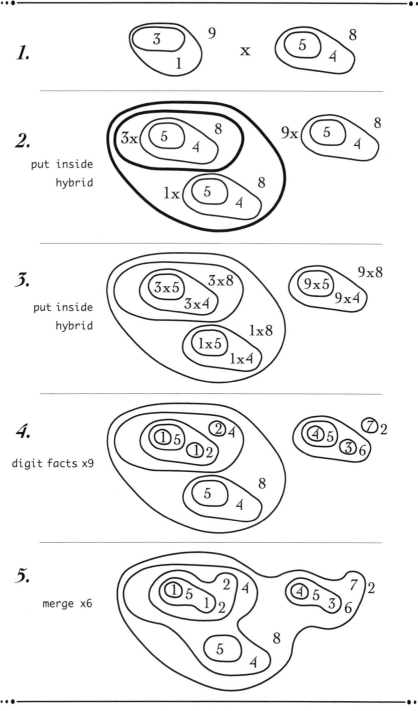

Figure 4-3: *Fluid-circle dialect:* 319x548, *frames 1-5*

Dynamics

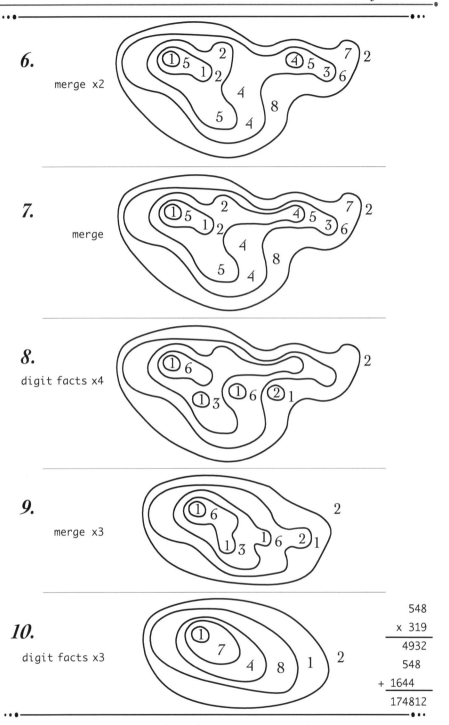

Figure 4-4: *Fluid-circle dialect:* 319×548, *frames 6-10*

Chapter 4

4.3 An Iconic Calculator

The stop-action animations of circle arithmetic in the previous section were constructed by hand. The boundary arithmetic calculator automates the display of the process of calculation. It illustrates the look and feel of container arithmetic in a dynamic responsive context.[4] The calculator display is in a **box dialect**, identical to the circle dialect but with straight bounding edges rather than bounding curves.

Animation: iconicmath.com/calculators/iconiccalculator/

This webpage introduces the **Iconic Calculator** and its functionality. There are three supporting pages, one for each calculator mode. Each page shows approximately the same calculations. The calculator implementation includes addition of positive and negative numbers; multiplication and division are not yet included.

Binary: iconicmath.com/calculators/binary/

The **Binary Calculator** page includes over twenty videos showing base-2 group, merge, reflect, and their interactive dynamics. The videos animate each transformation rule separately, and in various combinations that demonstrate parallel computation.

Decimal units: iconicmath.com/calculators/decimalunits/

The **Decimal Unit Calculator** page includes over twenty videos showing base-10 group, merge, reflect, and their interactive dynamics. Units are displayed as ensembles rather than digits.

Digits: iconicmath.com/calculators/digit/

The **Decimal Digit Calculator** page includes over thirty videos showing base-10 group, merge, reflect, and their interactive dynamics. Units are displayed as digits, with their interaction defined by the addition tables.

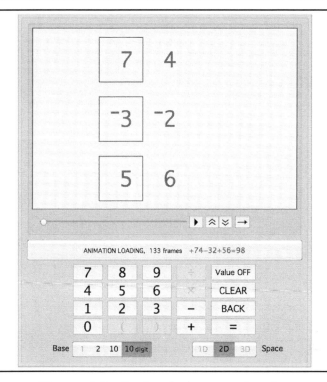

Figure 4-5: *Iconic calculator user interface*

Interface

The calculator interface is shown in Figure 4-5. Like a standard calculator, it has a keypad for entering desired computations. The significant difference is that the output not only shows the computed result, it also shows the boundary processes that generate that result. The interface displays the requested computation in box arithmetic, which then animates to show the smooth transformation of form from request to result. The dynamic display shows group, merge, and reflect, the same parallel reductions that are illustrated by parens arithmetic in Chapter 3.

Arithmetic expressions are typed into the keypad using conventional base-10 digits and the operation signs of arithmetic. A textual display bar above the keypad shows

Chapter 4

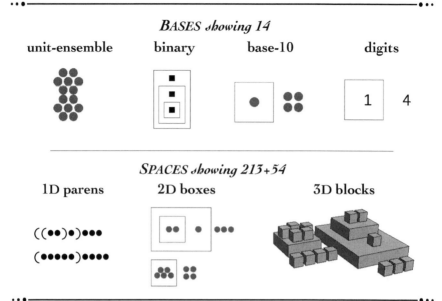

Figure 4-6: *Iconic calculator base and space modes*

the input, as well as animation readiness, status, and duration. The display bar also indicates any input errors such as a violation of input syntax. The display controls are standard video start/stop, slower/faster, and forward/reverse direction.

Modes

The calculator incorporates four computational **base modes**, shown in Figure 4-6 using the example of the number 14.

- base-1 shows unit-ensembles
- base-2 shows binary arithmetic
- base-10-units shows ensembles of size ten
- base-10-digit shows conventional digits

Figure 4-6 also includes examples of the display of the three **space modes** for the sum 213 + 54 prior to merging.

These space modes change the dimensionality of the display:

— one-dimensional parens
— two-dimensional boxes
— three-dimensional blocks

Animation

Figure 4-7 shows a selection of five animation frame sequences.[5] The five sequences in Figure 4-7 highlight specific aspects of the animation of arithmetic.

— *Sequence 1:* Put base-2 whole numbers.
213 + 54
— *Sequence 2:* Put base-10-units.
213 + 54
— *Sequence 3:* Base-10 digit group and merge.
777 + 164 + 689
— *Sequence 4:* Base-10 digit cancel.
20 − 41 − 136 + 2 + 258
— *Sequence 5:* Base-10 digit cancel and ungroup.
528 − 16 − 32

In base-2 and base-10 modes, the put operation aligns boundaries at the same level and then merges adjacent boundaries. In that sense, addition and merge are the same operation. In *Sequences 1* and *2* box forms move together until all boundaries converge to a single boundary, which is then removed to implement addition.

Sequence 3 shows parallel group and merge of base-10-units. Digits assemble in groups of ten in preparation for grouping. Some digits may subdivide, calling upon their underlying formal structure as unit-ensembles. *Sequence 4* shows base-10-digit parallel addition of five signed numbers. The mechanism of boundary addition is insensitive both to how many numbers are being added and to whether a number is positive or negative. Reflect deletes mixed polarity units in one-to-one correspondence.

Chapter 4

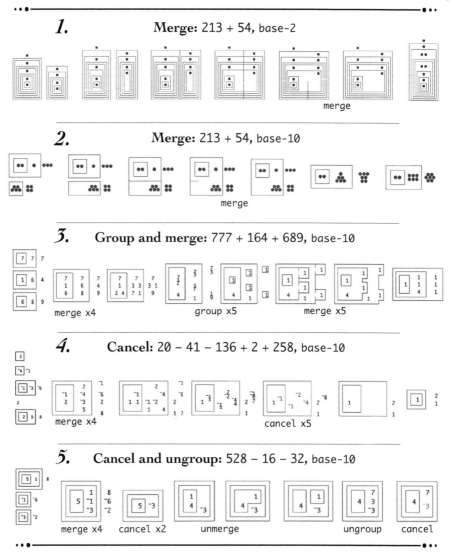

Figure 4-7: *Iconic calculator animation frames*

Sequence 5 illustrates a cancel and ungroup sequence. Digits are ungrouped to match with equal digits of opposite polarity. They move together and once overlapping, they fade into non-existence. When a digit polarity does not match that of the *deepest* unit, deeper units unmerge and ungroup to standardize the polarity of the entire form. Place-value subtraction calls this process *borrowing*.

Dynamics

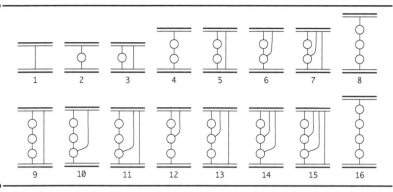

Figure 4-8: *Network numbers from* 1 *to* 16, base-2

4.4 Network Arithmetic

Network arithmetic is a relational dialect for which containers are objects, and containment relations are connections between objects. I've adopted a **gravitational convention** to display nesting in these networks. The ground is below, the containing object is *above* the contained object. Figure 4-8 shows the canonical base-2 network forms for the numbers 1 through 16.

A contains B

Animation: iconicmath.com/arithmetic/networks/

The **Networks** webpage shows a few video animations of network arithmetic, including the transformation rules and examples of addition and multiplication.

Network simplification follows the same rules as unit-ensemble arithmetic with binary depth-value. Networks also provide a *formal operational definition* for the concept of number. Rather than being inert objects that are operated upon, network numbers are active objects that compute themselves. Rather than building arithmetic from easily stated and relatively useless numeric objects coupled with computationally intensive operations, network numbers build arithmetic from contextually dynamic numbers combined with easily stated and relatively inert operators.

Chapter 4

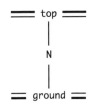

Reading

Network numbers have a **top** (output) and a **ground** (input). The ground is 1. Computation flows upward. Each **node** multiplies its input by the selected base. The top accumulates the network outputs.

To read a network number, start at the ground. Passing upward through a node doubles the value of the lower network input. When two subnetworks branch, their values add when entering a common upper node, usually at the top.

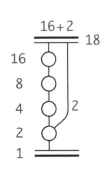

For example, to read the network form of the number 18, begin with 1 at the bottom. Passing through the first node yields 2. At this point a second path diverges, carrying the value 2 to the top. The original path carrying 2 continues upward, passing through three more doublers (4, 8, 16), finally sending the value 16 to the top. At the top, 16 is added to 2 to yield 18. The process is the same for base-10, except that each node multiplies its input by ten. In base-2 the value of the network in the sidebar is `10010`, which is the number 18 when expressed in base-10.

Standardization

The same network number has many different representations, that is, many different networks of connectivity. Networks with no nodes are tallies, each wire represents one unit. Without group and merge simplification, the top of a network bounds a unit-ensemble.

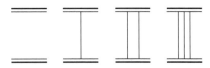

Standardization applies group and merge to minimize the effort of reading a network number. Figure 4-9 shows the standardization transformations in the network dialect.

Dynamics

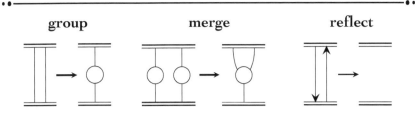

Figure 4-9: *Network transformations*

The group operation transforms a pair of simple connections into a binary multiplier represented by a node. The merge operation combines nodes that are connected to the same single lower node. Merge generalizes to any number of nodes at the same time, and is independent of the base. From the perspective of an implementation, *Merge is structure sharing*.

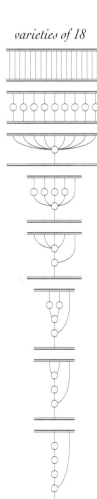

varieties of 18

The sidebar shows several different representations of the number 18. There is only one canonical minimal form, the form at the bottom of the page. The eight representations of 18, from top to bottom, also show the parallel standardization sequence from unit-ensemble to canonical form.

Standardization rules can be applied in parallel since every node is *top* to its lower neighbors and every node is *ground* to its upper neighbors. As well, the value of a network does not change when it is read upside-down, from bottom to top. **Self-similarity** is an important feature that allows all nodes to simplify and to compile values in parallel.

Addition

In boundary arithmetic, operations are implemented by putting networks together. Putting together side-by-side achieves addition, while putting together ground-to-top achieves multiplication. As illustrated in Figure 4-10, connecting boundaries is sufficient to perform either elementary operation.

Chapter 4

Figure 4-10: *Network add and multiply*

Figure 4-11 provides a dynamic example, 5 + 7. The two numbers being added are joined top-to-top and ground-to-ground on the first row. The second row shows the group and merge standardization sequence. Reordering is for visual convenience and is not a network transformation. The second and third images on the bottom row show the dynamic visualization of merging links sewing together and moving upward.

$$a + b + c = a\ b\ c$$

Multiplication

Multiplication is implemented by joining tops to grounds, displayed visually by stacking the network numbers. Removing the combined top/ground bar achieves multiplication of the upper form by the lower form. When there is more than one connection attached to the top or ground bar, the multiplied networks must be cross-connected. To **cross-connect**, join all the top links of the lower network to all the ground links of the upper network. The complete cross-connection of one network with another is a physical model of the Multiplicative Principle, that each link in one group is joined with (touches) each link in another group.

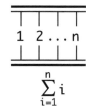

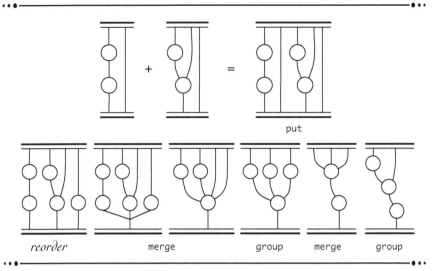

Figure 4-11: *Network addition:* 5+7

Any number of multiplications can be stacked concurrently. The order of stacking impacts the number of paths generated through the composite network. For binary networks, each network number will itself contribute a specific amount of ground complexity, either one or two upward connections. The number of top connections of each network is the count of 1s in the binary representation of that number.

The initial network form disentangles crossed links by replicating the upper subnetworks so that no node has more than one input. The number of replications is determined by the number of links connecting upward.

Figure 4-12 and Figure 4-13 show two animations, 7 x 5 and 5 x 7. In both animations, the network numbers are first stacked, then cross-connected. The cross-connected links generated by multiplication are in **factored form**. The lower rows of the two figures illustrate simplification after cross-connection.

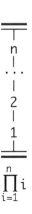

Chapter 4

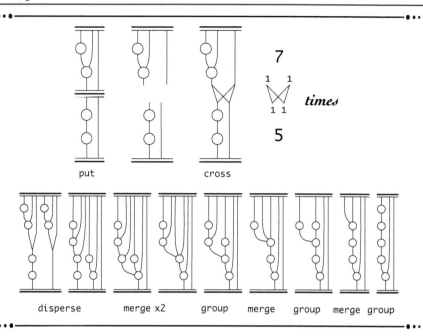

Figure 4-12: *Network multiplication:* 7x5

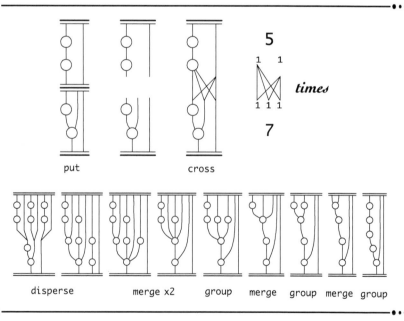

Figure 4-13: *Network multiplication:* 5x7

104

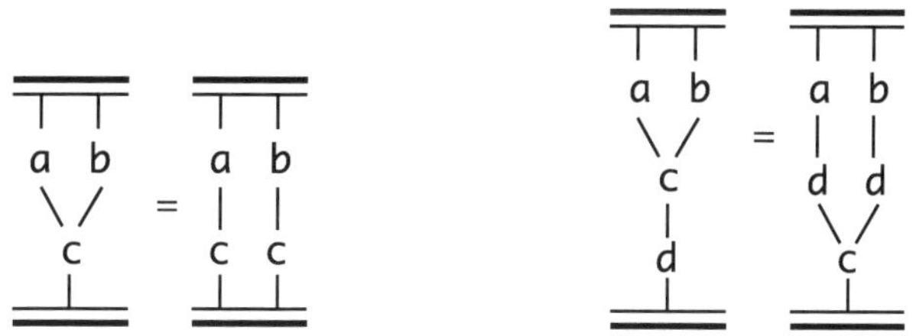

Dynamics

$$c \times (a + b) = (a + c) \times (a + c)$$

$$d \times c \times (a + b) = c \times ((d \times a) + (d \times b))$$

Figure 4-14: *Network arrangement*

Arrangement

By separating the cross-connections that result from multiplication-as-stacking, a factored form is converted into its polynomial form. Cross-connection is thus a succinct embodiment of what must be done to multiply polynomials. **Network arrangement** is illustrated in Figure 4-14. In both network equations, a lower node is unmerged.

Network arrangement has a more general form than can be conveniently expressed in symbolic text. In the network model, a form at any level can propagate through the network, exchanging positions with other forms. Propagation can be in either direction, toward top or toward bottom, since networks do not change the computed value when turned upside-down. Whenever the moving form encounters a branch, to continue to propagate, the form must be replicated once for each divergent path. This replication is named **dispersal** and corresponds to the distributive law of arithmetic. Dispersal appears in text as a combination of distribution and the commutativity of multiplication.[6]

Figure 4-15 shows the multiplication of two generic binomials. Dispersing the connections via unmerge generates the network equivalent of the polynomial expression.

Chapter 4

$$(a + b) \times (c + d) = (c + a) \times (c + b) \times (d + a) \times (d + b)$$

Figure 4-15: *Binomial multiplication as network arrangement*

Below, a binomial form is raised to the third power. The right-hand-side represents eight paths from ground to top. The eight possible paths correspond to the binomial structure of the conventional polynomial. One path passes through a three times; three paths pass through a twice and b once. Similarly, three paths pass through b twice and a once. One path passes through b three times.

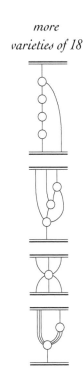

more varieties of 18

$$(a + b)^3 = 1a^3 + 3a^2b + 3ab^2 + 1b^3$$

Permitting multiple inputs and outputs for each node provides a diversity of new forms not usually encountered in textual notation. Four other examples of the network number 18 are illustrated in the sidebar. Network numbers are *not oriented*, rather they are bidirectional. Top and bottom can be reversed without impacting intention. Our gravitational bias provides an example of the Participation Principle: What we take as meaning depends on our choice of viewing perspective.

Implementation

Network diagrams can be interpreted as computer circuits. Connections are wires. Nodes are doublers implemented as shift-registers. When the ground is hot, the circuit outputs a binary number. Reading a network number is the same as computing its value.[7]

The computational effort of network numbers is in determining value rather than in adding or multiplying *per se*. Like unit-ensembles, the effort is in counting up the result, not implementing the operations. Addition is independent of the magnitude of the numbers being added; multiplication requires $\log_2 n$ effort, proportional to the length of the binary sequence that represents the number.

Negative numbers can be implemented by reversing top and ground. When tops and grounds are connected, networks then have a gradient. Links flowing upward (ground to top) are positive, links flowing downward (top to ground) are negative. Subtraction is the addition of networks with gradients. One additional standardization rule is necessary, the network analogy of the Unit Reflect rule. Characteristic of iconic notation, the canceling pair disappears completely. Reversing the direction of flow in a network is not practical as a way to implement physical circuitry, so negative numbers are not included in the implementation of the network dialect.

Decimal Networks

Although the network model was not designed for decimal computation, `base-10` decimal networks can be constructed using the same principles as binary networks. The primary difference is that nodes group ten links. There is also a pragmatic difference: the cost of looking up the interpretation of the digit symbols. We'll need the digit combination facts conventionally displayed as the addition and multiplication tables.

Chapter 4

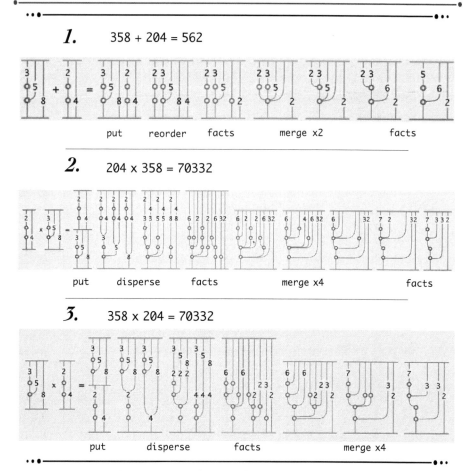

Figure 4-16: *Decimal-network addition and multiplication*

Figure 4-16 shows three sequences of addition and multiplication of decimal networks. *Sequence 1* shows frames from the animation of 358 + 204. First 358 and 204 are put together. This result is then standardized by grouping digits while applying addition number facts, and merging nodes as they are generated.

Sequences 2 and *3* show one multiplication in the two different commutative orderings. The first shows 204 x 358, while the second shows 358 x 204. In *Sequence 2* the two decimal networks are stacked and cross-connected. The result is then standardized. *Sequence 3* has

Dynamics

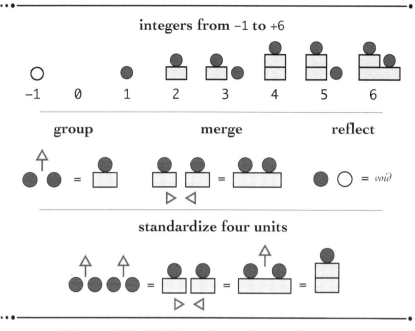

Figure 4-17: *Block numbers and operations*, base-2

fewer cross-connections. When 204 is stacked on top, six wires converge in the middle partition, creating six new paths. When 358 is stacked on top, only four new paths are created. Unlike base-2, the base-10 networks for commutative variants do not structurally converge until late in the standardization process.

4.5 Block Arithmetic

Finally, we'll cast the binary arithmetic of numbers into a three-dimensional manipulative dialect. Here, the illustrations are two-dimensional cross-sections. Units are represented by round disks, blocks by rectangles.

Animation: iconicmath.com/arithmetic/blocks/

The **Blocks** webpage describes the form and the transformation of block numbers, and does not include any videos or animations.

Chapter 4

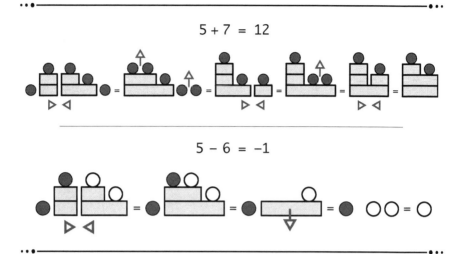

Figure 4-18: *Block addition and subtraction,* base-2: 5+7 *and* 5-6

Figure 4-17 shows base-2 block integers; the merge, group, and cancel operations expressed as movement of blocks; and an example of standardizing a block ensemble of four units. The group operation replaces two units by a block with a single unit on top. The merge operation joins blocks at the same level together. This dialect allows all depths to merge in parallel. Cancel, of course, deletes units with opposing polarity.

Figure 4-18 provides examples of addition and subtraction in block arithmetic. In the example of 5 – 6, a block without a unit on top doubles nothing, so is void-equivalent. When a negative unit is higher than a positive unit, the negative unit needs to be unmerged into the next lower level, which allows a pair of opposite polarity units to be canceled.

Figure 4-19 shows two examples of multiplication of block integers, both 7 x 5 and 5 x 7. In base-2, the structural difference between the two commutative varieties goes away early in the standardization process. After forms are multiplied by substitution, one parallel merge

Dynamics

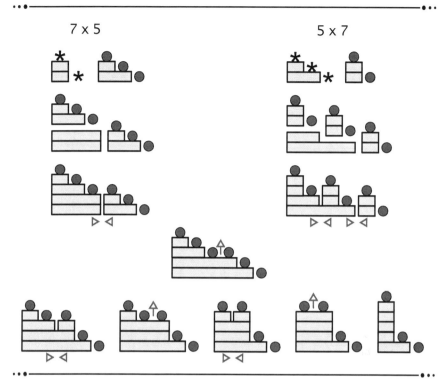

Figure 4-19: *Block multiplication,* base-2: 7x5 *and* 5x7

is sufficient to reduce both multiplied forms to the same structure. This structure is centered in the middle of the figure. The five subsequent transformations in the bottom row of the figure are identical for both multiplications.

The examples of block arithmetic thus far have all been in base-2. Figure 4-20 shows a base-10-digit multiplication, 319 x 548. This is the same example as that shown in the circle dialect in Figures 4-3 and 4-4. The block dialect shows parallelism particularly well. After the initial multiply-as-substitution, nine merges occur. This casts several digits together, which when added by grouping create three additional merges. The entire standardization occurs in four steps. In contrast, the circle dialect in Figures 4-3 and 4-4 shows six steps.

merging

2D circle

[[a]][[b]]
[[a] [b]]
[[a   b]]

3D block

[[a]][[b]]
[[a   b]]

Chapter 4

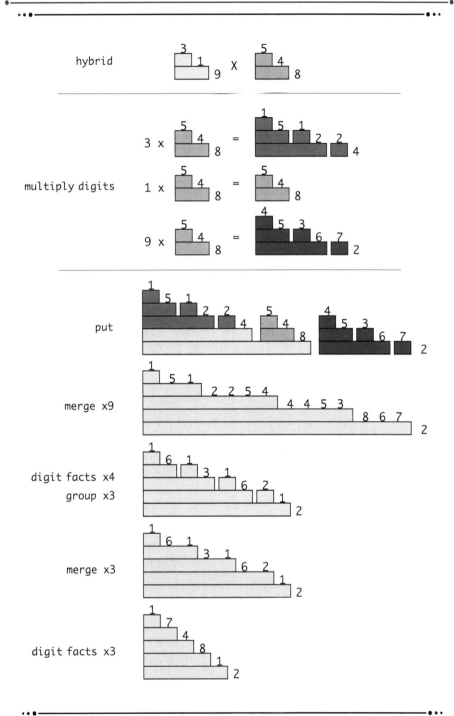

Figure 4-20: *Block multiplication,* base-10-digit: 319x548

Dynamics

Finally, the sidebar shows an example of block division of integers, 35 ÷ 5. Division is similar to multiplication, instead using ungroup and unmerge in alternation to deconstruct a block form sufficiently to identify patterns of blocks that match the form of the divisor, in this case 5, ((•))• in parens notation. From the deepest nesting, units successively ungroup until the largest substack of the divisor is matched. There is no specific deconstruction strategy for a particular divisor, other than the powerful ability to match *substacks* independently. In the sidebar example, the substack ((•)) is matched first, in the third frame. Deconstruction continues then on the subproblem of matching the remaining substack •.

Here's the sidebar division expressed step-by-step in parens notation.

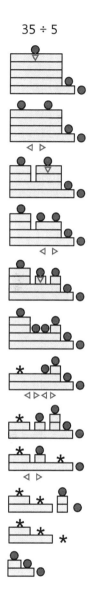

35 ÷ 5

```
35 ÷ 5   ☞   (((((  •  )   )))•)•
             ((((•         )))•)•
             ((((•))((  •  )))•)•
             ((((•))(•     •))•)•
             ((((•))(•)    (•))•)•
             ((((•))• •    (•))•)•
             ((   *     •  (•))•)•
             ((   *   )(•)((•))•)•
             ((   *   )(•)  *   )•
             ((   *   )    *)((•))•
             ((   *   )    *)  *
             ((   •   )    •)  •    ☞   7
```

4.6 Viewpoint

In the circle dialect we are looking down *from-above*, showing a topology of nested circles. The nested topology biases us to believe that we must cross over each boundary separately. The imposed step-wise linearity is an artifact of our viewing perspective. In the block dialect, we are looking *from-the-side*. Since we are able to see nesting levels as stacked rather than as embedded, parallel merge can occur across depth.

Chapter 4

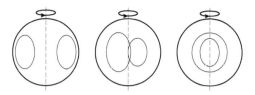

Viewing perspective changes topology.

Spencer Brown provides a clear example of the impact of viewing perspective on containment in the Notes to Chapter 12 of *Laws of Form*. Our concept of containment is supported visually only on a flat surface. Should we draw two circles on opposite sides of a sphere, rotation of the sphere can change the nesting configuration from ○○ to ◎. By modifying the viewing perspective, we can change operations that appear to be linearly nested at different depths to operations that are not nested and thus concurrent.

The first video on the iconicmath.com/arithmetic/networks/ page shows the binary network number 14 constructed in one, two and three dimensions. As the video changes our viewing perspective, the 1D textual form turns inside out, becoming nonsense. The 2D network shows its shallowness but returns to its original meaningful form. The 3D network is visually surprising in its robust maintenance of structure and meaning.

The **dimension of representation**, and thus the available viewing perspectives, interacts with what we take to be the meaning of numbers and of numeric arithmetic, an example of the Participation Principle.

One-dimensional symbols provide an impoverished point-of-view, forcing us to look from the outside, and thus from a purely cognitive perspective. We see meaning, *the way we interact with information*, as disembodied. Expressions without an interior encourage the poor psychological habits of arrogance (we can see everything), detachment from reality (we are external to context), fragmentation (we can understand by disassembling into pieces) and linearity (we can take only one step at a time).

Fields Medalist Timothy Gowers provides an illustration of the interaction of viewing perspective and interpretation.[8] I'll paraphrase it in boundaries. The assertion is $1 + 1 = 3$.

( ) is 1 from two different perspectives. Here we have been closing up the fractured boundary to see one enclosure. We can also not close it up, and count two characters and one gap between the characters. Now we'll put together (i.e. add) two bounded gaps

$$( ) + ( ) = ( ) ( )$$

Counting the *gaps*, one gap plus one gap yields three gaps. If we focus on characters as objects, then the above equation is $2 + 2 = 4$. If we focus on the juxtaposition relationships between pairs of characters, then $1 + 1 = 3$.

Two-dimensional icons provide a choice of perspective, but stack-up into a sequence when we do not exercise that perspective. The availability of a visual interior encourages introspection and participatory ownership. Knowledge based on experience becomes an option.

Three-dimensional models and forms connect directly with experience: visual, tactile, physical thinking.

4.7 Remarks

The transformation rules and sequences for all of the dynamic displays in the chapter are the same. The apparent variety is purely in the spatial representation of the same processes.

In education, higher dimensional experience has been replaced by symbolic myopia. Symbolic expressions, particularly in math, are both literally and syntactically non-sense. Symbolic knowledge costs more than confusion, more than disassociation from ourselves. It degrades understanding, performance, our bodies, our minds and our humanity. We loose the vital conceptual tools that we need to see the earth as a unified whole. We lose our senses.

Postsymbolic mathematics education is not a distortion that potentially undermines rigor. It is instead a vital force that might return sensory rigor and personal empowerment to the math classroom. There are two basic premises:

— Math should be obvious to all.
— We learn with our bodies.

With the visceral nature of unit-ensembles and depth-value in hand, we'll now move past rational arithmetic to numeric form in general by exploring James algebra.

Endnotes

1. **opening quote:** S. Wolfram (2007) *Mathematica Notation: Past and Future*. Online 8/2016 at http://www.stephenwolfram.com/publications/mathematical-notation-past-future/

2. **frames of the animation of 7 + 5 are shown in stop-action:** These animations were constructed with the 3D animation software Cheetah3D 6.0.

3. **when no further pattern-rules are triggered:** The fluid boundary representation on a flat surface still suffers from an artifact of too many apparent successive merge operations. Outer boundaries must merge prior to inner boundaries. The three-dimensional block dialect described later in the chapter allows concurrent merging of all levels of nesting. Nevertheless a collection of parallel changes may generate a pattern that itself triggers a transformation rule, resulting in a sequence of merge operations.

4. **the look and feel of container arithmetic in a dynamic responsive context:** The boundary arithmetic calculator was written in the programming language Mathematica, using standard techniques to compile and display

animations. Mathematica provides an elegant and versatile language with comprehensive built-in functionality. For example, the frame and buttons of the calculator interface shown in Figure 4-5 are objects that need only placement within a display and connection to their functionality. It was however quite challenging to assure smooth and consistent animation of boxes and numbers for *all possible combinations* of integers.

The animation engine consists of about 100 pages of formatted code, primarily due not to the complexity of animation, but to the diversity of options provided by the calculator. This project was never finished. The current calculator offers two-dimensional animation only, with stubs to accommodate extension to one-dimensional parens forms and three-dimensional block forms. For these extensions, only the display objects need modification. The computational steps, of course, are the same for any display mode. The calculator does include base-2, base-10 using unit-ensembles, and base-10 using conventional digits. It also includes pop-up evaluation flags that show the conventional values of what is happening at any given depth of boundary nesting.

There remain three significant unfinished sub-projects. Multiplication- and division-as-substitution are not implemented. Here the extension of functionality is straight-forward, but integrating the smooth display of substitutions is not. Second, conventional nested arithmetic forms such as 3 x (2 + 4) are not implemented. This imposes little challenge once multiplication is implemented, since the conventional expression is converted to boundary form at the input interface. These boundary forms do not need the delineated nesting of sub-expressions that disambiguate textual expressions. Finally, since the engine is prototype code, compiling the display animations is very slow. The display ANIMATION LOADING carries a delay of more than a minute in some cases.

5. **a selection of five animation frame sequences:** Base-1 mode unit-ensembles are not illustrated. For addition, the unit-ensembles move closer together. For subtraction, units of each polarity pair one-to-one and are deleted by cancel.

6. **a combination of distribution and the commutativity of multiplication:** From our learned text-based perspective, dispersal is a combination of two algebraic rules. From the network perspective, moving nodes across links is natural, while making the distinction that two different things are

occurring is artificial. We have associated the concept of multiplication with that of stacking, which makes any movement through the network invariant under the interpretation of multiplication. That is, multiplication is the fundamental operation of network arithmetic. Addition is secondary, appearing as the divergence and convergence of multiple links. A minimal network form has the least links, which corresponds to factored rather than polynomial textual forms. Minimality of links corresponds to minimality of nodes. In particular, we can consider cross-connection that removes a central node (the stacked bottom/top node) as increasing the *complexity* of a network form. Our textual and cultural preference for polynomial forms is an artifact of considering addition to be more fundamental than multiplication, and fractured textual forms to be more fundamental than integrated spatial forms. In James algebra, we will see a spatial arithmetic that treats exponentiation as fundamental.

7. **the same as computing its value:** A binary digit is a string of ones and zeros. For an eight-bit number, the network top would expect up to eight inputs (for the number 255). The network functions as a binary encoder. A hot ground is read as the byte 00000001. Each individual node functions like a binary shift register. The top reports the state of the byte. That is, the output of the circuit is the binary encoding of its computed value. These are fixed rather than general purpose circuits. Each circuit corresponds to a specific binary number.

8. **the interaction of viewing perspective and interpretation:** T. Gowers (circa 2005) Does Mathematics Need a Philosophy? Online 1/17 at https://www.dpmms.cam.ac.uk/~wtg10/philosophy.html

Chapter 5

Structure

Everything should be made as simple as possible, but not simpler.[1]
— Albert Einstein (1977)

This chapter is both a convenient summary of the James formal system collected into one place, and an advanced look at what is to come. It provides a succinct, structural description of the pattern rules that define James algebra. Chapter 6 next includes an introduction to the conceptual structure of these forms.

While learning, it is often beneficial to see in advance an overview of the upcoming content. This approach is called *advanced organization*, showing new concepts first so that when they are encountered later, in perhaps an unfamiliar context, the learner will not be taken aback by novelty. Cognitively, an advanced organizer allows the subconscious to process the new ideas and to lay a mental scaffolding that accommodates change. Our bodies do instantly understand the difference between inside and outside. Our skin is a very compelling (and built-in) model of a boundary. It's our cognition that needs reminding. I personally prefer the surprise of new ideas. You however might prefer to know what to expect.

Chapter 5

boundary type	James form	☞ interpretation
round	(A)	e^A
square	[A]	$\ln A$
angle	<A>	$-A$

Figure 5-1: *James boundaries and an interpretation*

5.1 James Algebra

There are **three types** of James containers, represented in linear text by three different shapes of delimiting brackets. Figure 5-1 shows these bracket forms as well as a possible interpretation for conventional arithmetic. When empty, the three container types can be interpreted as the three fundamental concepts of arithmetic: 1, ∞ and 0. They can also be read as operations: power, logarithm, and inverse.

Herein, James forms are presented from both the nested container perspective and from the numeric interpretation perspective. Neither perspective depends upon the other. The listings at the end of the chapter show permitted useful boundary transformations and connect the invariance that each supports to known expressions within conventional arithmetic and algebra.

some patterns of containment

To ameliorate the initial unfamiliarity, you may want to think of the notation for container forms as a foreign language. The goal is to see it as a clearer way to think about elementary arithmetic. We are also beginning at the very beginning, in conceptual territory so simple as to appear unfamiliar.

In mathematics, the greatest degree of self-evidence is usually not to be found quite at the beginning, but at some later point; hence the early deductions, until they reach this point, give reasons rather for believing the premises because true consequences follow from them, than for believing the consequences because they follow from the premises.[2]

Above, Whitehead and Russell are referring to deductive logic, but their comment applies just as well to an algebraic system. This chapter includes the entire content of James algebra, so that we may see all together the algebraic axioms and the structural consequences that follow.

Patterns and Principles

Many of the **concepts** of James algebra are succinctly listed on the Concepts page that follows this section. This conceptual structure rests upon a body of **general principles** related to boundary mathematics.

— The single underlying concept is *distinction*.

— Mathematics is the *experience* of abstraction.

— Experience is not a recording. Representation is not a reality.

*our body is
our interface*

— To participate in abstraction is to partition space, to construct a boundary.

— Boundaries both separate and connect.

— Representation and meaning are different sides of the same boundary.

In James algebra structurally different forms of containment are defined to be equal by pattern equations. **Forms** are configurations of boundaries. The structure of a form can be changed only by following specified pattern rules. **Axioms** are transformation rules that are permitted as design decisions. **Theorems** are convenient transformations that follow directly from the axioms. **Frames** are notational structures that allow a conceptual organization of types of forms. **Maps**, or **interpretations**, are ways to convert James forms into conventional expressions within the arithmetic and algebra of numbers. Some James forms can be interpreted as imaginary numbers, some can be interpreted as being non-numeric.

Arithmetic

Here are the iconic Principles that define elementary arithmetic:

- **Existence:** something is different from nothing.
- **Accumulation:** parts do not condense.
- **Additive:** a sum looks like its parts.
- **Multiplicative:** each part of one touches each part of the other.
- **Hume's:** equality is one-to-one correspondence.

Axiomatic Style

A mathematical system must include some undefined ideas from which other formalized ideas are constructed. Whitehead and Russell:

> It is to some extent optional what ideas we take as undefined in mathematics; the motives guiding our choice will be (1) to make the number of undefined ideas as small as possible, (2) as between two systems in which the number is equal, to choose the one which seems simpler and easier.[3]

We begin then as did Spencer Brown, with one concept, that of **distinction**.[4] We represent the distinction by a boundary with a clearly delineated inside and outside. If you like, we begin by assuming the Jordan curve theorem as perceptually obvious.

Axioms are structural starting points, the first ground. There is an unlimited variety of axiom sets. The few interesting ones are those that provide some sort of power: more elegant concepts, greater understanding, learnability, philosophical appeal, perceptual obviousness, or importantly, a clear map to well known formal structures like logic, numbers, and sets. You might notice that there are very few **definitions**. From the structural perspective, containers have no inherent meaning other than their ability to contain. Definitions are abbreviations. Again, Whitehead and Russell:

> ..."definition" does not appear among our primitive ideas, because the definitions are no part of our subject, but are, strictly speaking, mere typographical conveniences.[5]

Meaning is off-loaded onto an interpretation, so that we may read containment structures as physical forms, as collections of nested boxes. All that is required is the algebraic tool of an equal sign and the ability to substitute equals for equals. The interpretation is dragged along with the valid transformations. If we must, we can assume that the definition of "=" is *is-confused-with*.[6] In Chapter 7 we will identify equality as *permitted structural transformation*. This is in contrast to the usual interpretation of = as invariance of numeric value.

Notation

Our notational zoo includes only four types of creature.

- **containers** represented by delimiting brackets, with empty containers serving as constants.
- **variable letters** that stand in place of an arbitrary container with arbitrary contents.
- the **equal sign** which identifies both identities and permissible pattern substitutions. Its twin, the **not-equal sign** identifies perceptually obvious difference.[7] The double-arrow ⇔ identifies equality between entire equations.
- various **abbreviations** and **meta-symbols** that stand in place of arrangements that otherwise would be awkward to represent. The two types include the meta-concepts ⇒, ⇔, *indeterminate*, and *void*, and the finite list abbreviations ... and ..ₙ...

zoo

( ) [ ] < >

A a B b

= ≠

⇒ ⇔

...

..N..

void

bracket	name	use	chapters
		GENERAL	
{ }	set delimiter	conventional sets	
( )	shell	value-neutral outermost	2, 6-10, 14
⟨ ⟩	logic boundary	logic, not numeric	10, 15
		UNIT-ENSEMBLES, DEPTH-VALUE	
( )	parens	depth-value group	3-4
< >	angle	negative ensemble	2
⟦ ⟧	double-struck shell	substitution operator	2-4, 14
⦅ ⦆	double-struck round	depth-value base	11-12
		JAMES ALGEBRA	
o, ( )	round	numeric, exponential	6-13
[ ]	square	non-numeric, logarithmic	6-13
< >	angle	reflection, inverse	6, 10-13

Figure 5-2: *List of typographical delimiters*

5.2 Remarks

Figure 5-2 lists all of the typographical brackets used in this volume. They fall into three distinct categories. The general delimiters are not formally part of James algebra; they are in the meta-language. Depth-value delimiters are described in Chapters 2 through 4. The bracket system used in the rest of the volume is limited to the three James boundary forms: *round*, *square*, and *angle*. The empty round bracket has two representations, o and ( ), for typographical convenience. To reiterate, the "shape" names are arbitrary and have no connection to geometry.

The overall **motivation** is to learn a new and *quite different* way of thinking and to apply that thinking to elementary arithmetic. What we discover along the way is that conventional arithmetic appears to be an accumulation of design decisions that, taken as a whole, lack conceptual coherence. Over thousands of years, we have stumbled our way into an arithmetic that works, but like all evolutionary processes, the assembly of parts is rife with unnecessary and redundant appendages. We teach this conceptual jumble to our children and as a consequence they too continue to stumble through elementary arithmetic, most leaving school loathing mathematics.

The three pages that follow the Concepts page show all of the pattern transformations and interpretations included in this volume: the axioms, theorems, frames, and maps. The two final pages show the transformation patterns introduced in Volume II and Volume III. To take the bull by the horns, how on Earth can any human wade through the apparently cryptic representations that follow? By learning to identify some simple patterns.

Endnotes

1. **opening quote:** A. Einstein (1977) *Reader's Digest* October 1977.

2. **because they follow from the premises:** A. Whitehead & B. Russell (1910) *Principia Mathematica* Preface p.v.

3. **to choose the one which seems simpler and easier:** Whitehead & Russell, p.91.

4. **with one concept, that of distinction:** Spencer Brown, *Laws of Form*, p.1.

5. **but are, strictly speaking, mere typographical conveniences:** Whitehead & Russell, p.11.

6. **the definition of "=" is is-confused-with:** This perceptual perspective follows Spencer Brown's informal definition of the equal sign, p.69.

7. **the not-equal sign identifies perceptually obvious difference:** The asymmetry between = and ≠ is surprising. If two forms are not equal, we must be able to see the difference. If two forms are equal, it may not be immediately apparent since they may look different. Axioms, then, identify forms for which we cannot trust our perceptions directly. Axioms are designed confusion. The equal-sign unifies what we might otherwise believe to be different.

Chapter 5

Concepts

VOID
Void has no properties. (Nothing is not something.)
Form is either not nothing or an illusion.
Void-equivalent forms may vary in structure but not in relevance.
Void-equivalent forms are syntactically inert and semantically irrelevant.

CONTAINERS
Containers represent distinctions.
Everything is a container.
There is only one relation, contains.
Empty containers are units.
Containers are both object and process.

STRUCTURE
Forms are patterns of containment.
Valid forms can be constructed physically.
Forms can be represented in many multi-dimensional notations.
Containers support nesting and not sequence.
Containers are not limited to a specific capacity (no arity).
The contents of any container are mutually independent.

AXIOMS
Axioms subdivide existent forms into discrete groups.
Axioms define the forms that are void-equivalent.
All canonical forms are unequal.

EQUALITY
Containers with equal contents are equal.
Equals can be substituted for equals.
Removing identical outer boundaries maintains equality.
Removing equal contents maintains equality.
Equality is quantized dynamically by transformation steps.
Forms change meaning only when they cross a boundary.

ARITHMETIC
To count is to identify, categorize, indicate, fuse and label.
Addition is putting forms into the same container.
Multiplication is putting square forms into a round container.
Exponential and logarithmic bases are defined by the interpretation.
Inverses are represented by the same boundary in different contexts.

Axioms

ARITHMETIC

		page
( ) ≠ *void*	**existence**	168
( ) ( ) ≠ ( )	**unit accumulation**	170
([ ]) = [( )] = *void*	**void inversion**	184
<( )> ( ) = *void*	**unit reflection**	46

ALGEBRA

([A]) = [(A)] = A	**inversion** *enfold* ⇌ *clarify*	184
(A [B C]) = (A [B]) (A [C])	**arrangement** *collect* ⇌ *disperse*	193
A <A> = *void*	**reflection** *create* ⇌ *cancel*	241

Theorems

FRAME

(A [ ]) = *void*	**dominion** *emit* ⇌ *absorb*	242
([A][o]) = A	**indication** *unmark* ⇌ *mark*	218
([A][o..$_N$..o]) = A..$_N$..A	**replication** *replicate* ⇌ *tally*	219

REFLECT

<<A>> = A	**involution** *wrap* ⇌ *unwrap*	241
<A><B> = <A B>	**separation** *split* ⇌ *join*	241
<A <B>> = <A> B	**reaction**	242
(A [<B>]) = <(A [B])>	**promotion**	244
(A <[<B>]>) = <(A <[B]>)>	**promotion** *demote* ⇌ *promote*	

Chapter 5

Frames

([ ]) = *void*	**void**
([A]) = A	**inversion**
(A [ ]) = *void*	**dominion**
(A [B C]) = (A [B]) (A [C])	**arrangement**
([A][o]) = A	**indication**
([A][N]) = A..$_N$..A	**cardinality**
(A [B])	**magnitude**
(o [N])	**unit magnitude**
(<o>[N])	**decimal**
(J [A]) = <A>	**J-conversion**

Ensembles

(a|b|c) = (a b c)	**fuse**
〚A • E〛 = 〚E • A〛	**commute**
A <A> = *void*	**reflect**

Depth-value

•..$_N$..• = (•)	**group**
(a)(b) = (a b)	**merge**
((N)) = ([N] o)	**depth-value**

Logic

ARITHMETIC

〈 〉〈 〉 = 〈 〉	**calling**
〈〈 〉〉 = *void*	**crossing**

ALGEBRA

〈A 〈 〉〉 = *void*	**dominion**
〈〈A〉〉 = A	**involution**
A 〈A B〉 = A 〈B〉	**pervasion**

Structure

Maps

expression	☞	James form

UNITS

1	( )
0	< >
−1	<( )>
−∞	[ ]
$\log_\# -1$	[<( )>]

INVERSE

A	A
−A	<A>
1 ÷ A	(<[A]>)

ARITHMETIC

A + B	A B
A − B	A <B>
A × B	([A] [B])
A ÷ B	([A]<[B]>)

BASE

B^A	(([[B]] [A]))
B^{-A}	(([[B]] [<A>]))
$B^{1/A}$	(([[B]] <[A]>))
$\log_B A$	(<[[B]]>[[A]])

EMBEDDED BASE

#	(o)
$\#^A$	(A)
$\log_\# A$	[A]

PARALLEL

counting	1 + ... + 1	o ... o
addition	A + ... + Z	A ... Z
multiplication	A × ... × Z	([A]...[Z])
fraction	(A ×...× M)÷(N ×...× Z)	([A]...[M]<[N]...[Z]>)

Chapter 5

Volume II
Equations

A = B ⇔	A <B> = *void* = <A> B		**reflection bridge**
A = B ⇔	(A) = (B)		**compose context**
A = B ⇔	A C = B C	C ≠ [ ]	**compose content**
			decompose ⇌ *compose*
(A) = B ⇔	A = [B]		**equality inversion**
			cover ⇌ *cover*
A C = B ⇔	A = B <C>	C ≠ [ ]	**equality reflection**
			move ⇌ *move*

Two Boundary

<A>	= ([A])	**two-boundary angle**
[A]	= [([A])]	**two-boundary square**

Volume III
Non-numeric

AXIOMS

[ ] [ ] ⇒ [ ]	**unification**
<[ ]> <[ ]> ⇒ <[ ]>	**unification**
	→ *unify*
[ ] <[ ]> ⇒ *indeterminate*	**indeterminacy**

THEOREMS

(A <[ ]>) = (<[ ]>)	**dominion II**
	emit ⇌ *absorb*
[ ] = ([A][[ ]])	**square replication**
<[ ]> = ([A][<[ ]>])	
[ ] = (J [<[ ]>])	**square unit**
<[ ]> = (J [[ ]])	

INTERPRETATIVE AXIOM *and* THEOREMS

(<[ ]>) = <[ ]> = [<[ ]>]	**infinite interpretation**
[ ] = <[[ ]]>	**infinity**
<[ ]> = [[ ]] = [[[ ]]]	
(<[[ ]]>) ≠ *void*	**infinitesimal**

Volume III
J Patterns

THEOREMS

J = [<o>]	definition of J
<A> = (J [A])	J-conversion
J J = *void*	J-void object
[<(J)>] = *void*	J-void process
([J][2]) = *void*	J-void tally
J = <J>	J-self-inverse
[<(A)>] = A J	J-transparency
[J] = J [J]	J-absorption
A (J [A]) = *void*	J-occlusion

PARITY

N *even*	([J][N]) = ([J][<N>]) = *void*	J-parity
N *odd*	([J][N]) = ([J][<N>]) = J	
N *even*	J ([J][N]) = J	J-parity whole
N *odd*	J ([J][N]) = *void*	
N *even*	J ([J]<[N]>) = ([J]<[N]>)	J-parity part
N *odd*	J ([J]<[N]>) = *void*	

J FRAMES

(J [A]) = <A>	J-conversion
(J [<A>]) = A	J-involution
(J <[A]>) = <(<[A]>)>	J-angle
(J [J]) = J	J-self

COMPLEX

i ☞	(J/2 [o]) = (J/2)	*form of* i
π ☞	(J/2 [J])	*form of* π
a + bi ☞	a (J/2 [b])	*form of complex number*

Chapter 5

Chapter 6

Perspective

*Notation...is extremely important in mathematics.
A seemingly modest change of notation may suggest
a radical shift in viewpoint.*[1]
— *Barry Mazur (2003)*

James algebra is named after Jeffrey James. He and I developed the algebra as his 1993 Master's Thesis at University of Washington, *A Calculus of Number Based on Spatial Forms*. This work has not been published previously. Most of the results in Chapter 7 through Chapter 12 are included in Jeff's thesis. James numbers take their inspiration from Charles Sanders Pierce, who introduced boundary logic at the turn of the 20th century; from *Laws of Form*, the seminal work of the late George Spencer Brown; and from the work of Professor Louis Kauffman at the University of Illinois at Chicago.

James algebra is a radical reconceptualization of how we represent and think about conventional numeric operations (+, −, ×, ÷, ^, √, log). Like the unit-ensembles described in Chapter 2, James forms are *additive*. Forms are added together by putting them together into the same container.[2] Unlike unit-ensembles, multiplication is represented by a specific configuration of boundaries, rather than as the substitution operation. The advantage of this approach, especially during transformation, is that multiplication can be treated as a static pattern. By revoking the commutativity of substitution, we return match-and-substitute to its position as the only method of transformation, a generic mechanism that converts axioms into theorems and tools.

The axioms of modern algebra evolved in tandem with linear typography and with sequential (causal) thinking. The axiomatic structure of algebra and set theory do not capture the *essence* of what numbers are nor how they work since neither system rests upon the Additive Principle. We will presume that the manipulative use of numbers throughout history and the modern symbolic perspective on numbers both refer to the same numbers. Different constructions of the same concept provide alternative perspectives on that concept, perspectives that can enrich and generalize our current understanding. Our conventional perspective on arithmetic, the one currently taught in schools, is extremely useful for business and for scientific professionals, quite useful for sequential computers, and definitely a nuisance for educators, for students, for a great majority of Americans, and for parallel and concurrent computation.

6.1 Diagrammatic Math

Euler was first to propose a diagrammatic method of logic. His **Euler diagrams** associate embedded and overlapping circles with logic syllogisms. He explains:

Euler diagrams

> The foundation of all these forms is reduced to two principles, respecting the nature of *containing* and *contained*. I. Whatever is in the thing contained must likewise be in the thing containing, and II. Whatever is out of the containing must likewise be out of the contained.[3]

During the nineteenth century, non-Euclidean geometries were discovered and formalized. The previous two thousand years had established Euclid's geometry as the sacrosanct definition of mathematical rigor.⁴ But Euclid's parallel line postulate was too narrow, it worked only for flat surfaces. There are geometries, such as the surface of a sphere, for which the parallel postulate does not hold. In the nineteenth century, this trauma of discovery shook the mathematical world so fundamentally that the ancient Greek perspective of deriving mathematical knowledge from diagrams was completely abandoned, in favor of purely symbolic approaches. Herbert Simon: "Rigor, it was believed, called for reasoning to be formalized in symbols arranged in sentences and equations."⁵

Euclid
circa 450–350 BCE

Hilbert's program at the turn of the twentieth century set out to express mathematical reasoning in finite strings of symbols. Mathematical diagrams and other sensory/experiential forms were widely purged from rigorous mathematics. In particular Euler diagrams, Venn diagrams, Frege's deduction trees and C. S. Peirce's existential graphs, each of which has been shown to be sound, were all suppressed and largely forgotten.

Leonard Euler
1707–1783

John Venn
1834–1923

Charles Sanders Peirce
1839–1914

The **syntax/semantics barrier** is deeply implicated in the migration to linear structure. The meaning of words and strings of symbols became entirely separate from the words and symbols themselves. Understanding was buried underneath arbitrary obscurity. Symbols require augmentation, meaning must be added separately. Here's Larkin and Simon:

Gottlob Frege
1848–1925

> The fundamental difference between our diagrammatic and sentential representations is that the diagrammatic representation preserves exactly the information about the topological and geometric relations among the components of the problem, while the sentential representation does not.⁶

However, the Participation Principle reminds us,

Strings of symbols do impact meaning
by limiting how we think about what we are describing.

Chapter 6

Bertrand Russell
1872–1970

> There is, however, a complication about language as a method of representing a system, namely that words which mean relations are not themselves relations, but just as substantial or unsubstantial as other words. In this respect a map, for instance, is superior to language, since ... a relation is represented by a relation.

Bertrand Russell continues,

> I believe that this simple fact is at the bottom of the hopeless muddle which has prevailed in all schools of philosophy as to the nature of relations.[7]

James algebra builds all structure out of *icons, images and diagrams*, out of containers that support the visual relation of inside and outside. Logicians Barwise and Etchemendy:

> Diagrams are physical situations. They must be, since we can see them.... By choosing a representational scheme appropriately, so that the constraints on the diagrams have a good match with the constraints on the described situation, the diagram can generate a lot of information that the user never need infer. Rather, the user can simply read off the facts from the diagram as needed. This situation is in stark contrast to sentential inference, where even the most trivial consequence needs to be inferred explicitly.[8]

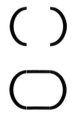

James algebra does not embrace linear, typographical communication. *James forms are spatial.* They are iconic rather than symbolic. The typographic representation used herein shows containers as string delimiters such as ( ), with the image of the container broken into a left and a right half. However, the sequencing and the fracturing of boundaries is an *artifact* of the way that our typewriters and our textual languages are constructed, and a potentially confusing distortion of the image of a container.[9]

Our theme is that numbers are sensory, diagrammatic, experiential. We do not need to obscure how arithmetic works with veils of symbols and tables of relations. Hiding meaning behind memorized convention both limits and distorts thought. Alfred North Whitehead: "By relieving the brain of all unnecessary work, a good notation sets it free to concentrate on more advanced problems."[10]

Boundary Algebra

First and foremost, James forms are **void-based**. *Void* is intended to look like what it means. It literally does not exist. In text the label *void* is a something, it's a word. Words are not nothing. They support reference and eventually communication. In contrast, nothing is not something, it does not support reference. *Void* is in our shared descriptive meta-language, however it is not part of the James system.

Void-equivalent structures and patterns are irrelevant to meaning. These forms exist solely in notation. Void-equivalent forms are illusions that arise from an empty page. Void-equivalence implies that

— *Absence* has no properties.

— Containers are *permeated* with void forms.

— *Void-equivalent forms* are background potentia. They can be freely created and deleted.

— *Empty* containers can be seen as units.

— Transformations *create and delete* structure.

James forms are **containers** with these properties

— Forms are *patterns* of containment.

— Forms are patterns of *physical* containers.

— A container is an *object* from the outside and a *process* from the inside.

— Contents are *mutually independent*.

— Concepts are *networks* of contains relations.

Chapter 6

formal arithmetic is putting stuff into the right boxes

A **calculus** consists of a notation for representing objects, a collection of permitted transformations and a collection of basic facts. We are representing physical containers by delimiting boundaries and basic facts by empty containers. Common use of numbers and arithmetic can be seen as putting things into containers and rearranging those containment relations by following the pattern rules defined by the three James structural axioms.

The physicality of containers means that we can viscerally interact with James forms. We can elect, for instance, to build James forms out of physical objects such as blocks, or out of physical enclosures such as rooms. The James axioms define the coordinated behavior of various patterns of containment. A sequence of structural transformations can be animated. The creation, deletion and rearrangement acts that constitute both proof and computation can be presented in videos as dynamic animations. Many transformations can happen at the same time since (other than containment) each container is independent of the others.

The inside of a container supports concurrent transformation of its contents, just like the inside of a theater full of people supports concurrent breathing. In that metaphor, all transactions are between a person and the air in the room, between content and context. There is no interaction between contents, no direct connections between people in the room. All may be immersed in (contained by) watching the movie. None are watching the other movie-goers.

In James algebra, there are no instances of counting, ordering or grouping because there is no imposition of structure other than that of containment. Importantly, only one axiom permits rearrangement of structure, the forte of string languages. The other two axioms (and most of the theorems) are void-based, they eliminate structure by erasure/deletion, by casting structure into *void*.

boundary	unit	interpretation	operator	interpretation
angle	< >	0	<A>	−A
round	( ), o	1	(A)	#^A
square	[ ]	−∞	[A]	$\log_\# A$

Figure 6-1: *James units and operations*

6.2 Container Types

James algebra uses three distinct types of containers to express numeric and non-numeric structure. Figure 6-1 shows the round boundary, the square boundary, and the angle boundary. In this volume, we'll stick to the interpretation of James forms in Figure 6-1.[11]

Boundary forms are icons. Pictorial forms trigger not only different conceptual models, they trigger different physiological processes. Transcription is therefore more than a cognitive shift, it implicates different perceptual systems and a different behavioral vocabulary.

The only relation within a boundary calculus is that of **containment**, a *minimal conceptual basis* consisting of one binary relation. The contains relation is quite general. When expressed within logic, containment can be interpreted as implies. When expressed as a network, containment is directly-connected-to. When expressed as a set, it's called is-a-member. When expressed as a number, it is successor. When expressed as a map, it's shares-a-common-border. Within the context of a pile of blocks, contains becomes supported-by. When seen as a family relationship, it is parent-of. When described as an abstract mathematical structure, it is a *rooted tree*. All of these metaphors share a collection of common characteristics that are concretized by the properties of physical containers. The fundamental *concept* underlying containment is **distinction**: a container distinguishes inside from outside.

Chapter 6

([A]) = [(A)] = A	**inversion**
	enfold ⇌ *clarify*
(A [B C]) = (A [B]) (A [C])	**arrangement**
	collect ⇌ *disperse*
A <A> = *void*	**reflection**
	create ⇌ *cancel*

Figure 6-2: *Pattern axioms of James algebra*

Volume II looks at the structure that the binary relations contains, implies, is-a-member, successor, and parent-of have in common: each makes a distinction between container and contained. This chapter provides an initial discussion of the mechanisms of the James pattern algebra of distinctions.

Pattern Axioms

Figure 6-2 shows the pattern axioms of James algebra, the transformations that are designed to define the behavior of the arithmetic and algebra of numbers.

Calculi based on an equal sign are called *algebras*. The algebraic style of boundary math includes maintaining equality by transforming containment relations that match clearly defined patterns. If a containment pattern does not match a rule, then it cannot be changed. More generally the **Axiomatic Principle** constrains what can be done within a formal system.

Axiomatic Principle
If it is not explicitly allowed, then it is forbidden.

Two of the three James axioms define the interplay between round and square boundaries, while the third defines the behavior of angle boundaries. It is convenient to be able to identify the direction of application of each axiom when describing computational steps. Figure 6-2 also provides these names. We will later develop several convenience theorems. The constructive demonstration of these theorems never strays far from the three simple axioms. All transformations of containment patterns are essentially simple, there are no particularly subtle or creative theorems in this volume.

The most important characteristic of these axioms is that two of them specify how to delete structure. Both implicate only one form (labeled A), so that they both require only simple pattern-matching. Remarkably, this leaves all of the complexity of numeric algebra isolated in one pattern transformation.

In general, a **variable** within an algebra stands in place of an *arbitrary* form. In James algebra, this idea is slightly more complex. Since the algebra is void-based, a variable might stand in place of nothing. Variables that are not void-equivalent stand in place of a single container and its contents. It is a violation of the structure of containment to have multiple forms standing in the same space without an outermost container. However, it is often typographically convenient to leave the outermost container implicit. We can display, for example, either (oooo) or ([oooo]) or oooo, with the understanding that the outermost container of oooo may be unwritten but it is certainly present.

explicit
([A B C])

implicit
A B C

arithmetic
([ ][ ])

Variables can be freely *deleted* to expose the arithmetic structure of a form. The algebra of James forms is thus strongly connected to its arithmetic.

algebra
([A][B])

Partial Ordering

Forms are specific patterns of containment. All the possible containment patterns constitute the **language** of James forms. The mathematical abstraction that comes closest to describing James forms is a partial ordering.

Chapter 6

A **partial ordering** is a graph consisting of nodes and links. The nodes are containers that are delineated by a boundary. The links are containment relations. A physical, or finite, strict partial ordering has these defining characteristics:

— There is a *top* node and a *bottom* node.
— There is a *direction*, every node is on a path between top and bottom.
— Links identify specific directional relations between nodes.
— Nodes bound links.

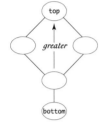

Within the theory of relations, a partial ordering has three characteristics

— **irreflexive:** no node is linked to itself
— **antisymmetric:** no node is above or below itself
— **transitive:** you can travel to distant nodes

One objective of Volume II is to look at these rather strange notions in detail. They fail to convey the intent of a containment pattern. Consider contains to be parent-of. You cannot be your own parent (irreflexive), and you can't be your parent's parent (antisymmetric). As well, your parents are not the parent-of your children (intransitive). Technically then containment is not a partial ordering because it is not transitive. There is a transitive concept that we might call contained-at-any-depth. In our example it would be the ancestor relation. But the deeper point is that using conventional abstractions based on sets does not particularly help us to think differently about the formal structure of distinction.

Interpretation

Figure 6-3 provides a quick introduction to the **interpretation** of boundary configurations that we will use. These patterns unify counting, adding, multiplying,

expression	☞	form
INVERSE		
A		A
−A		<A>
1 / A		(<[A]>)
ARITHMETIC		
A + B		A B
A − B		A <B>
A × B		([A] [B])
A / B		([A]<[B]>)
BASE		
B^A		(([[B]] [A]))
B^{-A}		(([[B]] [<A>]))
$B^{1/A}$		(([[B]] <[A]>))
$\log_B A$		(<[[B]]>[[A]])
EMBEDDED BASE		
#		(o)
$\#^A$		(A)
$\log_\# A$		[A]

Figure 6-3: *Algebraic operations to patterns of containment*

raising to a power, and the assortment of inverse operations. However, numbers and numeric operations are but one interpretation of what a container form might mean. When translating from one language to another (here, for example, between configurations of containers and conventional arithmetic), the more primitive, less redundant and therefore *foundational* language will have multiple alternative interpretations within the more sophisticated and complex language. Thus, the single boundary configuration A <B> can be read both as A + −B and as A − +B. Interpretation from a simpler foundation is one-to-many.

6.3 Multidimensional Form

A primary reason for going to all the trouble to learn this new sensual language is to learn new ways of thinking. It is not the concepts represented by the James language that are multi-dimensional, it is the language itself that has different dialects or "notations" expressed in different dimensions. One implication is that a series of transformations can be *animated*. Another implication is that many transformations can occur concurrently, all at the same time.[12]

When you stop to consider the rationality of symbolic representation, it becomes clear that symbols are highly discriminatory against our physical evolutionary heritage. The vast majority of the neurons in our brains are dedicated to managing the interface of our physical body with physical reality. Everybody lives in a body, only a very few of us live in the conceptual fantasy of the Platonic reality associated with mathematics.[13] Abstraction is of interest to only a small portion of a brain; the skills of abstraction are exceedingly difficult to teach. The symbolic math currently taught in schools expects us to abandon both sensation and experience in favor of unnatural cognitive acts. No wonder students find it difficult to learn this disembodied language.

symbolic

A contains B

iconic

concrete

experiential

Boundary languages are *visceral*. Interpretation will remain constant as the boundary representation is transcribed across dimensions, from 1D strings to 2D icons to 3D architectures to 4D temporal experiences. There is no abstract/concrete dichotomy, so that boundary languages are much easier to understand. No mind/body split, so boundary forms are much easier to tolerate. In contrast, string encoding cannot be experienced, it must be learned via memorization. Consequently string languages remain necessarily *cerebral*. **Mathematical nominalism** holds that mathematics is about objects that exist. Container languages provide nominalistic consistency by requiring that concepts too have a manifest form.

It is a distinct advantage to represent mathematical concepts across many different spatial formats, not only symbolically but also diagrammatically, physically and experientially. The printed page limits representation to symbolic and iconic forms, but by projecting volumetric forms onto paper, we can approximate concrete and experiential languages. The image of a box can elicit imagination of a box. Leibniz: "The best signs are images; and words, insofar as they are adequate, should represent images accurately."[14]

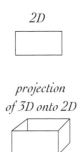

2D

projection of 3D onto 2D

Dialects

Containment relations themselves can be expressed not only as configurations of containers, but also as maps, networks and symbolic equations. Most string languages can also be expressed as spatial networks. A difference, though, is that a James icon embodies its operational semantics. In effect there is no distinction between form and intent. The container boundary is the only diagrammatic component we will need. It visually and computationally preserves the dependency of containment, which itself can be interpreted as nesting, sequence, stacking, connectivity and several other types of *physical* relationship between container and contained, as illustrated in Figures 6-4 and 6-5. Containers provide a built-in visualization of dependency, appealing for both form and interpretation to our hands and our eyes, rather than to our ears and vocal cords.

Figure 6-4 shows the James form of multiplication expressed as one-, two-, and three-dimensional containers.

— The **string** dialect is digital and encodable. The language consists of delimiters in fractured bracketing relationships with one another.
— The **bounded** dialect shows two-dimensional containment. The language consists of enclosures in nesting relationships with one another.

Chapter 6

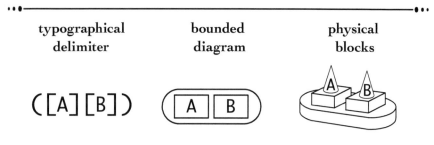

Figure 6-4: *One-, two- and three-dimensional forms of multiplication*

— The **block** dialect is manipulable. The language consists of physical objects in stacking relationships with one another.

Figure 6-5 shows James multiplication in some other spatial dialects of containment, including two-dimensional maps and paths, three-dimensional physical rooms and dimension-free networks.

— The **network** dialect is a traversable acyclic graph. The language consists of nodes and links.

— The **map** dialect is a traversable territory. The language consists of areas with shared borders.

— The **path** dialect shows border crossings that define the boundary form. The language consists of a single instance of each type of boundary, together with a directed path crossing the boundary archetypes.

— The **room** dialect is a three dimensional environment inhabited by a participant. The language consists of rooms and doors.

For examples of James *arithmetic* in each of these notations, delete the variables A and B in Figure 6-4 and Figure 6-5.

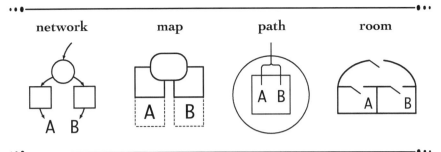

Figure 6-5: *More forms of multiplication*

Hybrid Notation

Hybrid notation mixes boundary structure with *special symbols* that stand in place of specific boundary forms. Usually these special symbols are reminiscent of conventional notations. Special symbols will be necessary in order to make the upcoming boundary forms of π, i, cosine, and the like easier to read. A hybrid notation makes unfamiliar forms a bit more readable. It is particularly handy when the same boundary pattern occurs multiple times within a form.

The archetypical special symbol is the variable A, which stands in place of any James form. *Naming* boundary structures is safe so long as unique names are assigned to unique patterns. And of course we can conveniently move back and forth between a name and the form that the name identifies. We can also engage in symbolic abbreviations for both typographical and reading convenience. We'll sometimes use natural numbers, for example, to stand in place of ensembles of units. It will be convenient to use the special symbol "5", for example, to stand in place of the form ooooo. Sometimes however it is conceptually important to *show* these ensembles explicitly.

o ☞ 1
oo ☞ 2
ooo ☞ 3
oooo ☞ 4
and so on

There are two abbreviations that we will be seeing a lot so they may as well be introduced now. The **hash mark**, #, indicates an *arbitrary base* for polynomial numbers and for exponents and logarithms. A new idea facilitated by James algebra is that exponents generally do not need a base for transformation to move forward. In an algebraic context, *which* number is being multiplied together many times (i.e. the *base*) often does not matter. Often an entire form is standardized for one base only. The base then is a global feature of the form rather than a local feature of each boundary, an example of the Communality Principle.

(A) ☞ base^A
[A] ☞ log$_{base}$ A

J represents another new idea, a new imaginary number, the **logarithm of negative one**. The special symbol J stands in place of the boundary configuration [<( )>]. J is the subject of Volume III and makes only a few appearances until then.

Object/Process

How should we think about −3, or 1−3, or 1/3, or √3, or log 3? These representations signify exact numbers, yet somehow we have embedded the operations of subtraction, division, root and logarithm into their notations. Can we do no better than to describe some numbers as operations on other numbers? Can combinations of units and operators legitimately be called numbers? This question can be phrased in the simplest of terms:

Is 1/3 a single number
or is it two numbers combined by an operation?

Does 1/3 explicitly identify an actual number? Yes, in conventional terms, it is a member of the class of rational numbers. Do the rational numbers cover the number line? No, no more than the natural numbers cover the number line. Does 1/3 explicitly identify a ratio or comparison? Yes, in conventional terms a fraction is a ratio arising from the comparison of the magnitude of two natural numbers. Here we compare 1 to 3. 1/3 identifies the proportion of the ensemble ••• that is •. So in a conventional sense, 1/3 is both one and two numbers, depending upon our purpose.

Perspective

The use the same notation to represent both concept and process is an example of the Participation Principle that is widespread in mathematics. *What a number means depends upon how you look at it.* This dual usage is extremely handy, since it permits an expert in mathematical abstraction to arbitrarily shift the interpretation of an expression between object and operation. 3 can be seen both as a number and as an operation on numbers, depending upon what is most convenient at the time. 3 can represent a count, or it can represent an act of replication. Not only does 1/3 behave as both a number and a comparison, the transformation rules for operating upon the structure 1/3 can differ depending upon its intended interpretation.[15] That's why expertise is needed to use conventional notations.

We are also asking an educational question: is this flexibility helpful to those learning how to manipulate numbers? Is the object/operator duality an essential aspect of what we mean by *number*, or is the duality an artifact of one particular way of looking at numbers? Might the distinction between object and operator be illusion, perhaps imposed by abstractions that are more complicated than is necessary? Are students and their teachers aware of the entanglement of things and processes? Why do fractions traumatize so many young students?

The object/operator duality of conventional numbers is reminiscent of the particle/wave duality of atomic particles. The founders of quantum mechanics became comfortable with the idea that an electron will behave both like a wave and like a particle, depending upon how we choose to observe it. What an electron *is* depends upon how we want it to appear. Similarly, what a number is depends upon what we want of it. Perhaps we should develop a conceptually simple arithmetic in which numbers do not take on the properties of operators. In which, for example, numbers add but do not multiply. Perhaps numbers can be simplified by limiting how they are permitted to behave.

Chapter 6

James numbers provide yet another perspective on the relationship between object and operator. Rather than embedding operations into the structure of numeric forms, James algebra uses *viewing perspective*, a feature of container-based math, to distinguish between object and operation. When we look at the outside of a container, ignoring its contents, then the container appears to be a very simple **object**, a container. When we look at the contents inside a container, the container itself is a **process**, a very simple process, contains. Viewed from the outside, a James container can represent a number. Looking at the inside shows us how that number is constructed. This is very similar to our use of function notation, where f(x) stands in place of the intended result of the process f acting upon the object x. But f(x) shows an inconsequential component of the inner structure, the arbitrary label used to access the functionality identified by the label f. The difference is that the contents of James containers are the entire inner structure.

outside/object

inside/process

6.4 Features

James numbers share these features with the other boundary systems:[16]

outside
inside
boundary

— There are significantly fewer abstract concepts.

— Everything is a containment relation.

— Void-equivalent forms can be ignored.

— Contents do not interact.

— Conventional operations condense into a few structural axioms.

— Proof and computation are achieved by pattern substitution and deletion.

— The algebraic theory of groups is not relevant.

Here is a slight expansion of each of these ideas within the context of James algebra.

Significantly fewer concepts

Perhaps the most challenging new idea is that many of the conventional *concepts* within arithmetic and algebra are unnecessary in order to conduct the conventional operations of arithmetic and algebra. Arithmetic can be far simpler than what we have been taught. The essential concepts of James algebra do *not* include zero, commutativity, associativity, arity, one-step-at-a-time processes, and symbolic strings. Instead there is a deep visceral unity across counting, addition, multiplication, power, subtraction, division, roots and logarithms.

Containment relations only

James algebra is based on *containment relations*. What is inside a specific container tells us what that container acts upon, or contains. Different concepts are expressed by three different containment relations, which we have represented as ( ), [ ] and < >. A numeric interpretation is not necessary and is not embodied within the axioms of the algebra.

Semantic use of *void*

The absence of a container has meaning. We can interpret absence as numeric zero. An empty container is a *unit*. Emptiness creates unity. Unity is absence of interior parts. The interior of every container, even those with contents, is pervaded by *void*. Void-equivalent forms are everywhere and in unlimited supply. *Bounded space* cannot and does not have properties, unlike the sequencing properties of the spaces between these words and the spaces between the characters of each word. *Void* is not synonymous with emptiness, it is more like the physical space that is underneath all physical objects. This leads to the fundamental **Principle of Void-equivalent Forms** (and of the *void* itself).

Void Equivalence
Void-equivalent forms are syntactically inert and semantically irrelevant.

In the numeric domain, void-equivalent forms are both blind to sign and blind to multiplicity.[17]

Independence of contents

There are no direct relations between contents. The only relation is between a container and each of its content forms. Another way to say this is that the void shared by contents of the same container has no properties and thus cannot support relations between contents. This explicit independence permits the implementation of extensive parallelism within forms. Concurrency is a native mode of thought, analogous to sight rather than to speech.

Algebraic operations condense into a few patterns

Three generic transformations on patterns of containment are taken as axioms. An interpretation of round-brackets as powers and square-brackets as logarithms permits ease of mapping James forms to conventional arithmetic. When nested, the two types of boundary maintain alternating exponential and logarithmic contexts which permit smooth transition between addition and multiplication.

Proof by pattern-matching and substitution

Axioms *statically* define patterns that are equivalent, and *dynamically* permit transformation between patterns. Structural axioms are implemented by matching a given structure to a permitted pattern and then replacing it by a certified equivalent structure. Two axioms identify void-equivalent forms, permitting deletion of structure. The third permits rearrangement. This makes computation and verification short and elegant.

Absence of the laws of algebra

Commutativity and associativity are interpreted as sequential concepts that are not relevant to a spatial, parallel form of arithmetic.[18] There is no ordering imposed on the contents of a container. Grouping is defined solely by containment. Unlike functions which specify a precise number of arguments, a container can accommodate any number of contents (arguments). There is no concept of *arity*. The 0 null object of addition and the 1 null object of multiplication are opposite sides of the same distinction, 1 from the outside and 0 from the inside.

Incidentally, if you are counting, we have eliminated three of the five fundamental properties of algebraic groups. From the James perspective only inverse and distribution are fundamental.[19]

6.5 Strategy

James algebra may not be initially friendly to the eye, primarily due to its unfamiliarity. It becomes friendly to the mind only after exerting some effort to achieve familiarity. Familiarity can be achieved cognitively or iconicly or physically. Do not expect to easily recognize a radical revision of the mathematics that has been standardized across the globe for a century. Do not expect to easily shed the perspective embodied within twelve years of grade school and high school mathematics courses. Or, if you choose, simply adopt the mind of a novice, someone who has never encountered a number greater than two.

Mathematical concepts themselves are defined by containment patterns. We do not have to abandon the familiar concepts of symbolic math, since carrying them into a reading of a James form at least provides comfort in familiarity. The adventure though is in finding other concepts, *iconic concepts*, overlaid and interacting with our familiar concept of number. It is these new perspectives that provide the motivation to explore. Could it be that the categories of numbers that we hold to be important (naturals, integers, rationals, irrationals, transcendentals, reals, imaginaries) have accreted over time without

Chapter 6

common varieties of numbers

0
1
43
−5
½
5/13
32/5
2 1/7
2.97
10110
√2
log 7
e
π
i
3+4i

fundamental organization? Might there not be an alternative classification of number that makes more sense? Similar to Conway numbers[20] (aka surreal numbers), this exploration is not yet pragmatic; the goal is simply to shed some new light on what it means to be a number. A deeper objective is to show that symbolic arithmetic as we know it is not fundamental, but rather just one of many formal approaches to magnitude.

The exposition of James numbers includes both definition and transformation, as well as guidance about how to use the new ideas. Some of the new iconic concepts are not about numbers at all. They are spatial rather than numeric. These same concepts also permit us to describe and to simplify conventional *logic*, thus achieving a unification of number and logic, of algebra and proof.[21] One small change in the rule that governs units can change arithmetic into logic. Visually:

natural numbers ( ) ( ) ≠ ( )

elementary logic ( ) ( ) = ( )

Let's first look at James numbers dispassionately, as just another way to think about magnitude. We will establish a map between James forms and conventional numbers. Later we will ask: if both systems describe the same thing, why do they appear to express such different concepts? How can the single idea of containment cover the diversity of ideas embodied in conventional arithmetic? More specifically, is the (bewildering to the uninitiated) array of number types in the sidebar at all reasonable?

6.6 Remarks

You may have noticed that Wikipedia articles and technical publications on mathematics are often impenetrable, both symbolically and conceptually. That's reasonable, mathematics is after all a highly technical field of study. We would not benefit here, however, by a purely mathematical approach since one of the primary design principles of boundary math is to help to make math more comprehensible to the non-professional. The strategy is to demonstrate formal techniques within a psychologically motivated and physically friendly system of communication. However, we *are* exploring a formal system, and formal systems inherently describe how a computer works rather than how an organic being thinks.

I've approached this description of James algebra as if we were learning a new language together. Well, a bit more than just a new language, also a new way to think about numbers, a new set of concepts. Comments in the text highlight significant differences between the boundary-based approach and the conventional symbolic approach to notation. A form looks like what it is intended to mean, although it does not necessarily look like how we might interpret it as a conventional number.

Boundary math does not seek novelty as a goal, it seeks simplicity. Most usually, the simplest path is both invisible and obvious. Mathematician Alexander Grothendieck:

> The very notion of a scheme has a childlike simplicity -- so simple, so humble in fact that no one before me had the audacity to take it seriously.[22]

The algebra of boundaries shares with Grothendieck the goal of simplicity, achieving it not by a higher level of abstraction of categories of mappings, but by embracing *only one thing*, the distinction.

Coming up, after introduction to the various James representations and transformations in Chapters 7 and 8, we'll look closely at how counting and arithmetic work in Chapter 9. We'll see the form of addition and multiplication as the interplay between two types of boundaries, ( ) and [ ]. And we'll look at the basic structural ideas underneath the natural numbers.

Chapter 6

In Chapter 10, we'll take the next big step to introduce the *generalized inverse* and a third type of boundary, < >. Angle boundaries come with a third axiom, so we end up with one axiom that defines each type of boundary, although ( ) and [ ] are co-defined. With James arithmetic in hand, finally in Chapter 11 we will be able look at the entire range of conventional numbers.

We'll generalize the developed tools to include rational expressions and other number types. Then we'll move arithmetic into the higher dimensions of spatial forms. Toward the end of the volume, we'll take a quick look at several other boundary arithmetics.

The [ ] unit in combination with angle-brackets creates a cacophony of non-numeric structures including the exotic and indeterminate expressions of conventional math such as negative infinity, infinitesimal, divide-by-zero, square-root-of-negative-one and logarithm-of-negative-one. Volume III retreats to the uninterpreted James arithmetic to examine the forms that correspond to these exotic expressions, viewing the development of theorems as design choices that come with both strong and weak points. This quasi-mathematical approach is thus a hybrid of rigor and realism. Volume III also focuses on one particular form, [<( )>], with very unusual properties.

Volume II and Volume III each contain a significant surprise. In Volume III it is a *new imaginary unit*, one that is more fundamental than $\sqrt{-1}$. In Volume II it is the disclosure that the James angle-bracket is only a *convenient shorthand abbreviation* that allows us to contrast negative and positive and thus remain in familiar cognitive territory where negative numbers are taken to exist. The numeric inverses too are configurations of only round- and square-brackets. In Volume II we'll see that James algebra has only two independent boundaries that are bonded together by one void-based axiom and one rearrangement axiom.

Endnotes

1. **opening quote:** B. Mazur (2003) *Imagining Numbers* p.163. Mazur continues on p.166: "It is also easy to underestimate the difficulties of comprehension that any change of notation presents."

2. **by putting them together into the same container:** Like unit-ensembles, depth-value notation is available to simplify large collections of James units. In Chapter 11 we will see depth-value notation arise naturally as a James form.

3. **out of the containing must likewise be out of the contained:** L. Euler (1802) Letter CIV Different Forms of Syllogisms, (2/21/1761), H. Hunter (trans.) *Letters of Euler* p.406.

4. **Euclid's geometry as the sacrosanct definition of mathematical rigor:** Euclid's *Elements* (circa 300 BCE) is a collection of mathematical proofs that was the second most published book (behind *The Bible*) in Western civilization for over 2000 years. To be a mathematician up until about 1850 was to memorize *The Elements*.

5. **to be formalized in symbols arranged in sentences and equations:** H. Simon, forward to B. Chandrasekaran, J. Glasgow, and N. Narayanan, (eds.) (1995) *Diagrammatic Reasoning* p.xi.

6. **while the sentential representation does not:** J. Larkin & H. Simon (1987) Why a Diagram is (sometimes) Worth Ten Thousand Words. In Chandrasekaran *et al*, p.696.

7. **in all schools of philosophy as to the nature of relations:** B. Russell (1923) Vagueness. *Australasian Journal of Philosophy and Psychology* (1) p.84–92.

8. **even the most trivial consequence needs to be inferred explicitly:** J. Barwise & J. Etchemendy (1996) Visual Information and Valid Reasoning. In G. Allwein & J. Barwise (eds.) *Logical Reasoning with Diagrams* p.23.

9. **a potentially confusing distortion of the image of a container:** The overwhelming majority of published analyses of containment address string-based concepts dominated by the right-half (called open) and the left-half (called close) of a fractured container. Algebraic representation is built upon

a stream of characters, whereas geometric representation occupies space. *Algebraic geometry* has all but banished geometric form from mathematics.

10. a good notation sets it free to concentrate on more advanced problems: A. Whitehead (1958) *An Introduction to Mathematics* p.39.

11. the interpretation of James forms in Figure 6-1: Multiplication-as-substitution, characteristic of unit-ensembles, will not be used in the following chapters.

12. many transformations can occur concurrently, all at the same time: Chapters 4 and 13 include several examples.

13. the conceptual fantasy of Platonic reality associated with mathematics: L. Bunt, P. Jones & J. Bedient (1976) *The Historical Roots of Elementary Mathematics* p.122:

> Most mathematicians accept the modern philosophical ideas that their axioms are logically arbitrary and that their theorems are about mental concepts. These mental concepts cannot be actually observed in the physical world. This view of the nature of mathematics can be traced back to the Greek philosopher Plato.

14. words, insofar as they are adequate, should represent images accurately: Leibniz to Tschirnhaus, end of 1679 (Math., IV, 481; Brief., I, 405), as quoted in L. Couturant *The Logic of Leibniz*, Ch. 4 footnote 93.

15. differ depending upon its intended interpretation: Treating the same symbolic expression in different ways depending upon its *external context* (i.e. its interpretation) can potentially invalidate the use of substitution, equality and identity, making the teaching of mathematics problematic. Symbolic expressions are treated as both clearly defined and ambiguous at the same time. An example of a context dependent system is written language. Both symbolic math and written text are languages. However in the prior sentence what we take as the meaning of the word "language" is determined by its applied mathematical or linguistic context. See E. Gray & D. Tall (1994) Duality, Ambiguity and Flexibility: A Proceptual View of Simple Arithmetic. *Journal for Research in Mathematics Education* **26**(2) p.115–141. Online 9/16 at http://homepages.warwick.ac.uk/staff/David.Tall/pdfs/dot1991h-gray-procept-pme.pdf

In a delightful and often quoted address to mathematics educators, Vladimir Arnold emphasizes that axioms are simply properties of transformations. This is quite similar to the approach taken here, that the pattern axioms identified in Chapter 5 are mechanisms that permit us to identify the absence of difference, i.e. the lack of a distinction.

V. Arnold (1998) On teaching mathematics. A. Goryunov (trans.) *Russian Math. Surveys* 53(1), p.229–236. Online 10/16 at http://pauli.uni-muenster.de/~munsteg/arnold.html and http://www.math.fsu.edu/~wxm/Arnold.htm Also available at several other academic sites.

16. **share these features with the other boundary systems:** Chapter 14 includes some of these alternative boundary systems. Unit-ensembles, spatial algebra and boundary logic are all described at iconicmath.com.

17. **void-equivalent forms are both blind to sign and blind to multiplicity:** *Void*, of course, has no properties including the absence of both polarity and replication. A void-equivalent form does have a typographical *presence*, however it is a technical error to think that a void-equivalent form is a specific form. Presence is essentially arbitrary and refers to *any* void-equivalent form. Hybrid forms such as ±([ ]) and 2 x ([ ]) tempt us to attribute sign and multiplicity to nothing at all.

18. **sequential concepts that are not relevant to a spatial, parallel form of arithmetic:** Peirce appears to be the first logician/mathematician to suggest that commutativity and associativity are secondary rather than axiomatic concepts.

> (4.374) Operations of commutation, like xy therefore yx, may be dispensed with by not recognizing any order of arrangement as significant. Associative transformations, like (xy)z therefore x(yz), which is a species of commutation, will be dispensed with in the same way; that is, by recognizing an equiparent as what it is, a symbol of an unordered set.

C.S. Peirce (1931-58) *Collected Papers of Charles Sanders Peirce*. Hartshorne, Weiss & Burks (eds.).

19. **only inverse and distribution are fundamental:** Volume II shows that even the concept of inverse is derivative. Network numbers in Chapter 4 and

Chapter 6

the James Arrangement axiom in Chapter 8 show that algebraic distribution is a special instance of a broader concept.

20. **Similar to Conway numbers:** J. Conway (1976) *On Numbers and Games*.

21. **achieving a unification of number and logic, of algebra and proof:** This unification has been a fundamental goal of work in boundary math since Spencer Brown.

22. **no one before me had the audacity to take it seriously:** A. Grothendieck, quoted in R. Hersh and V. John-Steiner (2011) *Loving + Hating Mathematics* p.116.

Chapter 7

Units

*Things are entirely what they appear to be —
and behind them ... there is nothing.*[1]
— *Jean-Paul Sartre (1938)*

James algebra incorporates three types of boundaries, represented here by three types of typographical delimiters. The **round boundary**, ( ), can be interpreted as the prototype numeric unit, One. The **square boundary**, [ ], is the prototype non-numeric unit, Negative Infinity. The **angle boundary**, < >, is the prototype void-equivalent unit, Zero.

*an empty container
defines a unit*

( )	☞	1
[ ]	☞	$-\infty$
< >	☞	0

Types of unit
numeric
non-numeric
void-equivalent

Natural numbers are ensembles of round units. Round boundaries (alternatively round containers, round-brackets) *accumulate*. In contrast, neither the square nor the angle boundary supports accumulation or multiplicity.

The empty round boundary is intended to look like one whole.[2] The emptiness inside is intended to look like nothing, i.e. the absence of quantity that we conventionally represent by 0. A James boundary can be interpreted as a

Chapter 7

function that operates on its content. With no content, the boundary is a constant. Without interpretation, James forms are simply *patterns of containment*.

We'll take now the first step toward void-based thinking. An empty container affords us the ability to refer indirectly to nothing at all. We cannot begin with nothing at all, since there would nothing to begin with. We'll use the meta-token *void* to refer to "nothing at all" within the textual commentary. If anything, the token *void* is a null index, a bridge that allows us not to be too confused by the absence of form. But *void* is not within James algebra, it is a crutch rather than an object. **Meta-language** talks *about* a formal system but the concepts and expressions within a meta-language are not *within* the system itself.[3] We cannot refer to *void* within the James algebra because there is no form with which to associate a reference. However, we can indirectly use *void* outside of the algebra to associate reference with the contents of an empty container. Within the algebra, an empty container itself provides indirect reference to its non-existent contents, while Inversion provides a void-equivalent form.

void has no properties

void inversion

$$([\,]) = void$$

In all cases, equations such as Void Inversion are hybrid and are more properly written without recourse to the meta-language.

$$([\,]) =$$

For convenience of description, I will not emphasize, in every sentence, that common words for numbers and operations are *interpretations* and are not necessarily properties of containment relations. Similarly, there are boundary concepts to enforce. An ensemble that lacks an outer container, for example, requires you, the reader, to exert care not to impose an operational interpretation on forms "sharing" an empty space. Similarly, we cannot *replace* absence by zero(s), an essentially impossible feat since, like physical space, *void* permeates all forms everywhere.

7.1 Nothing

The ancient Greeks found *nothing* to be a very controversial concept. The world, throughout all ancient and many modern cultures, was believed to originate from *something*, whether it be the primal ocean or ur-matter or birthing by the gods. Henning Genz recounts that the world was full, "there was no room for nothingness".[4] Democritus and the atomists found a use for empty space: it provides room for atoms to move. To maintain the primacy of the natural numbers, the Pythagoreans had to banish irrational numbers into the void. There is nothing between 1 and 2. Aristotle identified the void as the ground of distinction:

> The void distinguishes the natures of things, since it is the thing that separates and distinguishes the successive terms in a series. This happens in the first case of numbers; for the void distinguishes their nature.[5]

Specializing in the virtual, mathematics accommodates the idea that there must be something by filling in the potential void with dimensionless objects called **points**. Genz again:

> To the mathematician, space is nothing but an ensemble of points that are connected by some set of relations. In this interpretation, the laws of nature that apply to objects, including our probes, have nothing whatever to do with the properties of the space around them.[6]

Until recently, all mathematical spaces have been characterized by a set of points. Writing in 1969, Courant cautions

> But in modern mathematics the word "space" is used to denote any system of objects for which a notion of distance or neighborhood is defined...[7]

We must be careful to remember that absolute nothingness has throughout Western culture been associated with ultimate evil. "Nature abhors a vacuum" is attributed to

Chapter 7

Aristotle, and for more than a thousand years, his words were taken to be truth. Augustine equated Nothing with the Devil; it represented complete separation from God, loss and deprivation from all that was part of God, an ultimate state of sin, the very antithesis of a state of grace and the presence of God. *Nothing* represented the greatest evil.[8] During the Renaissance, the idea of *horror vacui* was inextricably entwined with Christian theology. It was not until Galileo that the concept of **looking to see what is** was invented, although he too asserted *"Natura abhorret vacuum."* The conceptual change introduced by the Western European Renaissance was to abandon the absolute knowledge of theology and acknowledge human ignorance.[9] We crossed from the complete fullness of objective perception into the possible emptiness of thought.

Zero

The token 0 stands in place of the absence of something. Throughout history, it has been viewed as a nexus of ambiguity. Theorist Anthony Wilden:

> Zero is not an absence, not nothing, not the sign of a thing, not a simple exclusion.... It is not an integer, but a meta-integer, a rule about integers and their relationships.[10]

Brian Rotman, in *Signifying Nothing*, accurately describes the semiotic conundrum of the 0 symbol.

> It is this double aspect of zero, as a sign inside the number system and as a meta-sign, a sign-about-signs outside it, that has allowed zero to serve as the site of ambiguity between an empty character...and a character of emptiness, a symbol that signifies nothing.[11]

Should we wish to interpret the space enclosed by an empty container, we might call it 0. But the use of 0 to identify nothing is misleading since it introduces superfluous symbolic properties. The symbol 0 has a specific location and a specific multiplicity; neither is a property of *void*. For example, in conventional notation it is possible to convey the idea of a multiplicity of "nothings":

a symbol cannot be nothing

$$0 + 0 + \ldots + 0 + 0$$

The construction of 0 in place of *void* introduces pure distortion by associating both a place and a multiplicity with a nothing that does not support location, cardinality, or polarity.[12]

Much Less

The nothing we begin with is much less. There is a nothing about which we cannot speak. We finesse this issue in the meta-language by calling it *void*. There is a different nothing that we *can* refer to, the emptiness inside an empty container, ( ). Speaking of the first destroys it. We avoid this destruction in speaking of the second by not speaking of it but instead speaking of its container.

Putting even the symbol 0 into an empty container renders that container no longer empty.[13] The complete absence of a representation of zero, although possibly mysterious, is easy to understand. We do not need to write down something in order to say nothing. Neither the place-holder role of zero nor the algebraic role of zero as the numeric null object are necessary for a definition of natural numbers. Zero is a symbolic patch over a poor design.

Paul Lockhart emphasizes that origin, unit of distance and orientation in space are choices that we impose upon an empty space and are not inherent in space (or *void*) itself.

> Space has no orientation, no natural unit, and no special place. There is no such thing as left or right, up or down, big or small, here or there, until we make these choices.[14]

Chapter 7

The properties we impose upon nothing are invariably associated with our *bodies*, putting physical perception at the foundation of geometric representation. Fundamentally scale, orientation and location refer to us. Mathematical generalization quickly constructs properties that are not expressible in physical terms. Mathematical abstraction quickly constructs the illusion that there is no mathematician. Regardless, the origin of both generalization and abstraction is physical experience.

Mathematics, Keith Devlin asserts, cannot be abstract.

> The brain can think only in real, concrete terms. We do not learn to think abstractly; we learn to make the abstract appear real. We do not extend into a new realm, we pull back into a familiar one.[15]

One

An empty round boundary encloses nothing and can be interpreted as the unit One. Within the language of boundary forms, we would say that the whole arises out of framing nothing. The conceptual foundation of *bounded nothing* provides a singular perspective from which to understand the formal system of numeric arithmetic. A boundary effectively partitions empty space into two identifiable locations that we call inside and outside. Which is which is purely an artifact of where we place our own viewing perspective. We culturally identify with "outside".

nothing is inside unity

Spencer Brown begins *Laws of Form* with this:

> *The theme of this book is that*
> *a universe comes into being*
> *when a space is severed or taken apart.*

It is important to realize that the universe is defined by the severing boundary, not by the apparent pair of spaces. Inside and outside are different perspectives on the same distinction. We cannot, of course, create copies of nothing. Here is Maturana and Varela:

( ) is a container, nothing more

A unity (entity, object) is brought forth by an act of distinction. Conversely, each time we refer to a unity in our descriptions, we are implying the operation of distinction that defines it and makes it possible.[16]

Since our fundamental numeric unit can contain other forms, it has both a static and a dynamic reading, as an object and as an operation. It's convenient to have two alternative representations of round boundaries. We'll use the token o to emphasize the empty round-bracket as a unit: an *object* that provides the numeric ground. We'll use ( ) to emphasize the round boundary as an operation: a *process* that transforms its contents.

$$( ) = o$$

What operation does the round boundary apply to Nothing in order to generate Unity? This question can be formulated as a conventional function. What operation (call it R) satisfies the functional form R(0) = 1? Whatever it is, the operator R cannot call upon any number (such as add-two, or divide-by-three), because we do not as yet have numbers. For that matter, we do not yet have *logic* to draw upon. We are developing the concept of Number from a perspective that does not consider the concepts of Logic or Sets or Multiplicity to be fundamental. We have *only* a Boundary around Nothing, together with an interpretation that takes ( ) as unity.

1 ( )
2 (( ))
3 ((( )))
...

One interpretation is that R is the successor function, +1. This interpretation uses nesting to mean accumulation, effectively squandering the representational freedom made available by nested containers. Another interpretation of R is the exponential function. ( ) can be interpreted as any base raised to the zero-ith power, which is conventionally defined to be 1.

1 +1(0)
2 +1(+1(0))
3 +1(+1(+1(0)))
...

$$( ) \mathrel{\text{☞}} 1 \qquad \#^0 = 1$$

We'll interpret a round-bracket (A) to be a **power operator** that raises an arbitrary base to the power of the boundary's contents, $\#^A$.

$1^0 = 3^0 = 629^0 =$
$(-31)^0 = (2/3)^0 =$
$x^0 = 1$

Chapter 7

The round boundary could be called an exponential boundary or a power boundary, but these names predispose us to think of James algebra as numeric. *Round* is a convenient label suggesting nothing about the interpretation of the boundary. James algebra makes no distinction between the object *one* and the function *raise-to-the-zero-power*. Either of these interpretations is available at any time by electing to read the round boundary from the outside as an object or from the inside as a process.

Existence

If we begin with nothing, where does the *equal sign* = come from? This delicate question is explored more fully in Volume II. Equality is the absence of distinction. The base case is

$$\textit{void} = \textit{void}$$

which can be more elegantly stated without the crutch of meta-language as

void equality
$$\boxed{\quad = \quad}$$

Equality constructs two indistinguishable perspectives. Since *void* cannot be replicated or partitioned, these two perspectives are created by a distinction. Equality is that distinction, one that does not distinguish inside from outside. *Equality is a distinction that does not distinguish.*

$$= \neq \textit{void}$$

The **Existence Principle** provides the definition of the inequality sign, ≠. This sign is not a test of unequal values. The concept of "value" comes much later, after an initial distinction has been made, as does the concept of logical negation. ≠ is a declaration of distinguishability.

existence
$$\boxed{\quad (\,) \neq \textit{void} \quad}$$

Existence
We can distinguish something from nothing.

Existence is not the comparison of a form to the absence of a form. It is rather a comparison between the outside of an empty container and the inside. Outside, each and every container represents *not nothing*. From the outside we associate empty containers with unity, with that which has no parts.

≠ *is crossing a boundary*

Inequality identifies a change in viewing perspective: looking from the outside vs looking from the inside. More abstractly, the boundary definition of the ≠ sign is that it identifies a **difference**. ≠ declares that crossing a boundary (whether physically or perceptually) constructs a difference. Difference, in turn, is the construction of a distinction.

cross inward

crossing changes meaning

The idea of constructing a distinction is visceral. We see an *object* from the outside. When we cross the boundary to the inside, we can no longer see the object. Instead we see an *environment*. In an empty environment we see nothing.[17] Existence is created (or observed if you prefer) by the act of crossing a distinction from inside to outside. Mathematician Rudy Rucker:

cross outward

> What sensation is really about is making distinctions.... It is this making of distinctions that leads to the world of number.[18]

We can eliminate the *void* on the right of Existence by making the outermost container explicit.

The shell-bracket shifts our perspective on ≠ from a comparison of the existence or non-existence of an object to a comparison of the presence or absence of contents. Very literally, shifting from implicit to explicit framing changes our reading style from active to passive, while it changes our interpretation from process to object, from transform to form.[19]

7.2 Natural Numbers

Throughout history, natural numbers have been considered to be **tallies**, collections of identical forms that we've called *ensembles*. Figure 7-1 shows James natural numbers as ensembles of round units within a common container. Ensembles require an outermost container, the top-level if you will, to assemble the constituent unit forms. This so as not to confuse us into thinking that units are able to autonomously group themselves into the things we call numbers.

In Figure 7-1 the **shell-bracket**, ( ), provides a **neutral outermost boundary** and serves as a reminder that ensembles are always bounded. The shell-bracket is another meta-token. It is not a James form and it has no conventional interpretation. If anything, ( ) makes explicit the page margins that frame all text. Just like typographical margins, the shell-bracket can be either implicit or explicit, depending upon the focus of the reader.[20]

Later we will see transformation patterns that allow us to place any bracket of our choice as outermost, without changing the intention of a form. To express that the *shell bracket is void-equivalent*, we can assert formally

([A]) = A
[(A)] = A
<<A>> = A

$$(\ ) = void$$

Accumulation

Natural numbers are grounded by the concept of distinction. The concept of *repetition* generates the plethora of distinctions. The construction of whole numbers proceeds in the same manner as it did for millennia, via the accumulation of units.[21] Like all tally systems, an essential property of James numbers is that forms add simply by being placed together. Formal grounding for the Additive Principle is provided by the Accumulation axiom of James arithmetic:

round units do not combine

unit accumulation $(\ )(\ ) \neq (\ )$

Units

		outermost container	
		implicit	*explicit*
0	☞		( )
1	☞	o	o
2	☞	o o	(o o)
3	☞	o o o	(o o o)
N	☞	o..ₙ..o	(o..ₙ..o)

Figure 7-1: *Integers to James units*

To make the outermost container explicit, we can write

$$(\;(\;)(\;)\;) \neq (\;)$$

The **Principle of Accumulation** defines the additive property of round units, and is indeed the heart of the concept of a natural number.

Accumulation
Parts accumulate rather than condense.

A consequence of accumulation is that all James forms that can be interpreted as numbers will include o in the deepest level of their canonical form.

Cardinality

There is a *prior* accumulation event, in which one unit accumulates alone. Prior to accumulation, there is **existence**, a concept that is well-defined in a void-based system. The self-similar step from Existence to Unit Accumulation asserts, via inequality, a most elementary principle of arithmetic: **one-to-one correspondence**. Finite ensembles cannot be put into one-to-one correspondence unless they are perceptually and operationally identical. Since all units within an ensemble are themselves identical, the only characteristic available to distinguish different

Chapter 7

ensembles is the failure of correspondence via one-to-one matching. The base case for failure of one-to-one correspondence is Existence. When units remain after one-to-one pairing of the contents of two ensembles, we can conclude the ensembles are different since a unit does not equal nothing.

cardinality does not depend upon counting

We can call the multiplicity of forms grouped together by a common boundary the **cardinality** of that boundary. The concept of cardinality applies to any container. Cardinality requires only the absence of interaction between contents. When the contents of a container are identical replicas, it is natural to consider cardinality to be a natural number.

Accumulation does not depend upon labeling each unit. *Determining equality does not depend upon counting.* Later we will use Accumulation to disambiguate a fundamental confusion introduced by teaching children to point at objects and chant 1, 2, 3…. When "three" is declared, is the object being pointed at labeled by the word "three"? Or is it the third object pointed at in sequence? Or could any object be counted as "number three"? Is counting related at all to sequential pointing? Indeed, is sequential labeling by sequential whole numbers related at all to the question of how many?

Base-free Exponents

specific bases are not necessary

The operation of *raising-to-a-power* can be **base-free** just as an algebraic variable can be number-free when it applies to all numbers. We will interpret the round boundary as raising # to the power of its contents

$$(A) \quad \Rightarrow \quad \#^A$$

For consistency, the square boundary then is interpreted as the base-free logarithm of its contents.[22]

$$[A] \quad \Rightarrow \quad \log_\# A$$

Units

The *hash mark*, #, is a meta-symbol that is neither part of the James notation nor part of the interpretation. # acts like a universal variable standing in place of an *arbitrary base*. In our interpretation, boundaries can stand in place of exponential and logarithmic expressions. In contrast, transformation within James forms is independent of the concept of a base. In effect, # is a convenience that permits the *interpretation* to remain uncommitted to a value of a base.

In the case of the empty round boundary, ( ), it does not matter what the base is because under conventional interpretation any base still yields 1. We can embed any particular base into our interpretation of the empty round boundary. To explore transcendental numbers, for example, it is useful to embed the base e.

( )	☞	$10^0 = 1$	*embedded base-10*
( )	☞	$e^0 = 1$	*embedded base-e*
( )	☞	$\#^0 = 1$	*base-free*

The central idea is that the concept of base itself is an *interpretative add-on*, not part of the container algebra. To *interpret* a round boundary as a conventional expression, we will need to provide both the operation (raise-to-a-power) and the base (raise-*this*-to-a-power). There is a James form that we'll explore later that incorporates the idea of an explicit base.

B^A ☞
`((([B]][A]))`

The hash-mark is a special symbol for the form (( )). The value of a **double round boundary** is the value of the base assigned to a single round boundary.

( )	☞	$\#^0 = 1$
(( ))	☞	$\#^1 = \#$

The concept is strange and bears repeating. Within the exponential interpretation, (( )) is not a particular number, rather it is a specific form that we can safely impose any numeric value upon, so long as we apply that value

173

base-10 example
(oo) ☞ $10^2 = 100$

consistently during interpretation. Boundary transformation is not numeric computation, it is a way to achieve conventional computation without referring to specific numbers themselves. The hash-mark is not a variable, not an arbitrary number; it is a meta-token that supports an *arbitrary interpretation*. Thus the *numeric* value of #^A is indeterminate until # is fixed by assignment to a concrete number. This is not much different than raising a conventional variable to a power, except that within the expression x^A, the x is expected to be quantified either universally or existentially. Within the James algebra, in contrast, we are not raising-to-the-power-of A, we are simply putting A into a round container. No type of quantification is implied.

There is a conventional way to understand the implications of an arbitrary base. Any choice of base is just a constant factor away from any other choice of base. The constant of proportionality, $1/\log_B N$, relates the chosen base B to an arbitrary base N.

$$\log_N A = 1/\log_B N \times \log_B A$$

The constant of proportionality is simply a scaling factor, the same type of scaling factor that we use to convert between units of measurement, such as measuring in kilometers rather than in miles. Educator Paul Lockhart:

> Choosing a base for our exponents is exactly the same as choosing a measuring unit — essentially we are measuring the size of a number by how many decimal digits it has.... If what you want is to convert multiplication to addition, then any base is as good as any other.[23]

The hash-mark defers binding a base until it is provided by an application context. Similarly, the size of a unit is arbitrary, a perspective supported by thinking of units as a measure of distance between 0 and what we call "1". In a void-based system, however, the "size of" ( ) is the "distance" between non-existence and existence.

7.3 Unification

As a type of unit, [ ] must have different additive properties than ( ), otherwise the two would be indistinguishable. Round-brackets provide a tally via Accumulation, which explicitly forbids deleting replicated units. In contrast, [ ] *dominates* rather than accumulates. We can express this as an iconic equation, one that reduces apparent multiplicity to unity. The Unification axiom defines the behavior of square-brackets.

$$[\,]\,[\,] \Rightarrow [\,]$$

unification
$\rightarrow$ *unify*

Although this axiom is an equality, it is only used in the direction of simplification. The square-bracket does not support replication. It is a semantic error to believe that more than a singular [ ] can exist, given its pervasive domination of form. Prior to the concept of counting, we cannot say: "Two of them is the same as one of them," a perspective driven perhaps by an expectation that we are dealing with numeric forms. We must say "The appearance of more than a singularity is an illusion." There is no structural way to generate imaginary replicas, the existence of [ ] is total. This circumstance is formalized in Chapter 10. The path dialect in Chapter 13 provides a visualization.

Square-brackets are blind to multiplicity, making them clearly non-numeric. Strictly speaking, [ ] does not *absorb* the forms within its environment (no forms in the same space interact). More accurately, [ ] completely fills its container, so that the outer container can support no other contents. From the perspective of the outer container, Unification asserts that there is only one empty square-bracket whereas Accumulation asserts that every round-bracket makes a difference.

Unification provides an interpretation of conventionally extreme numeric cases, particularly when 0 is implicated. It is the interpretation of [ ] that may be cause for concern, because we immediately encounter forms that cannot be interpreted as numbers. Our current interpretation reads square-brackets as taking a logarithm. An *empty* square-bracket represents the logarithm of zero, but what is log 0? The logarithm of 1 is 0. The logarithm of a number less than one, say $1/\#^A$, is negative, since

$$\log_\# 1/\#^A = \log_\# \#^{-A} = -A \log_\# \# = -A$$

As A grows larger, $\#^{-A}$ grows smaller while the logarithm of $\#^{-A}$ becomes a larger and larger negative number. The conventional limit of $\#^{-A}$ as A goes to infinity is 0, while the limit of $\log_\#$ N as N goes to 0 is $-\infty$.

$$\log_\# (\lim_{A \to \infty} \#^{-A}) = \log_\# 0 = -\infty \quad \text{☞} \quad [\,]$$

We can interpret [ ] as *negative infinity*, whenever the *void* within the square-bracket is interpreted as 0. Infinity is not a number and must be carefully manipulated in conventional notation using the technique of limits. Interpreting an empty square-bracket as negative infinity leads us into the exotic arithmetic of infinities, a topic explored in Volume III. We can avoid these difficulties for now by the simple expedient of leaving [ ] *undefined for interpretation*. However, [ ] is a valid form to manipulate within James algebra, given of course that we follow the pattern rules specified by the James axioms.

7.4 Remarks

We've built units by framing nothing and letting forms accumulate into whole number tallies. The nesting of forms within forms is sufficient to express the variety of conventional finite numeric expressions. At the foundation of the endeavor is a deep respect that *void* does not support relations. A direct consequence is the loss of ability to distinguish object from process. What a boundary is depends upon how we view it.

In his critique of the concept of number, French philosopher Alain Badiou asks

> Must we stop with Frege, Dedekind, Cantor or Peano? Hasn't anything *happened* in the thinking of number?... Isn't *another idea* of number necessary, in order for us to turn thought back against the despotism of number...has mathematics simply stood by silently during the comprehensive social integration of number, over which it formerly had a monopoly?[24]

Badiou continues

> The three challenges to which a modern doctrine of number must address itself are those of the infinite, of zero, and of the absence of any grounding by the One.[25]

Since we have eliminated zero and have exiled infinity to the fantasy realms, we need only to follow the consequences of building solely from the singular idea of distinction. The Existence Principle provides grounding for One.

To communicate our exploration, we have found need for several meta-tokens that bridge between boundary and conventional concepts.

- *void*   cannot be expressed within the form
- #   is an arbitrary numeric base
- ( )   assures an outermost container
- ...   is a finite list of unspecified length
- ..N..   is a finite list of length N

Chapter 8 next introduces two pattern transformations that define equality of form. These axioms support a consistent interpretation that anchors James patterns to the rules of conventional arithmetic. One axiom is void-equivalent, it specializes in deleting structure. The other limits rearrangement. Oscillation between the two, together with the limitation of structure solely to containment relations, is all that is needed to demonstrate the algebraic (group theoretic) structure of the arithmetic of natural numbers.

Chapter 7

Endnotes

1. **opening quote:** J.P. Sartre (1938) L. Alexander (trans.) *Nausea*.

2. **intended to look like one whole**: The typographical ( ) is a spatial enclosure with *implicit* top and bottom boundaries. Read ( ) as one container, not as two characters.

3. **are not within the system itself:** Formal systems embrace rigorous rules, but formality is so limited by its axioms and transformation rules that it is better understood as silicon computation rather than organic thought. In all mathematical systems, it is necessary to engage in a separate language, outside of the formal system, that is descriptive of the intentions of that system. Mathematical notation itself is like a map rather than a territory. It's useful when you need it, but the content of math is psychological, not symbolic. Spencer Brown (*Laws of Form* p.v.) adds yet another layer of subtlety:

> Mathematics is a way of saying less and less about more and more. A mathematical text is thus not an end in itself, but a key to a world beyond the compass of ordinary description.

Later (p.xx) in reference to the formal rules themselves, he observes:

> I found it easier to acquire an access to the laws themselves than to determine a satisfactory way of communicating them. In general, the more universal the law, the more it seems to resist expression in any particular mode.

4. **there was no room for nothingness:** H. Genz (1999) *Nothingness: The science of empty space* p.33.

In his second chapter, *Nothing, Nobody, Nowhere, Never: Philosophical, linguistic, and religious ideas on nothingness*, Genz provides a comprehensive scholarly summary of the early history of *void*.

5. **for the void distinguishes their nature:** This quote is from the Wikipedia page *Pythagoreanism* (online 2/16). It is attributed to Aristotle in *Protrepticus*, as reconstructed by D. Hutchenson and M. Johnson (2015).

6. **have nothing whatever to do with the properties of the space around them:** Genz, p.44.

Units

7. for which a notion of distance or neighborhood is defined: R. Courant & H. Robbins (1969) *What is Mathematics?* p.250.

These **manifolds** have the **metric property** which allows different points to be related as locations, whether they be points on a number line, or infinite dimensional points in a Hilbert space, or points that support a boundary in a rubber-sheet topological space. A manifold is defined by the types of relations between its points. But recently, category theory has expanded the geometric sub-structure of mathematical spaces from points to collections of more complex objects such as tangent lines (an example of a **sheaf**) and even to collections of all possible algebraic relations supported by algebraic objects (an example of a **topos** or **scheme**). The general idea is to build structures not from sets but from mappings between undefined objects.

8. Nothing represented the greatest evil: J. Barrow (2000) *The Book of Nothing* p.68. Barrow's book contains an excellent description of the origins of number systems. Nothing did have some Renaissance fans. "Among the great things which are found among us the existence of Nothing is the greatest." DaVinci, *The Notebook*, translated and edited by E. Macurdy (1954) p.61.

9. abandon the absolute knowledge of theology and acknowledge human ignorance: Y. Harari (2015) *Sapiens: A brief history of humankind* Chapter 14.

10. a rule about integers and their relationships: A. Wilden (1972) *System and Structure: Essays in communication and exchange* p.188.

11. a symbol that signifies nothing: B. Rotman (1987) *Signifying Nothing: The semiotics of zero* p.13.

12. a nothing that does not support location, cardinality, or polarity: The difference in mathematical perspective is to shift from an external coordinate system grounded by an arbitrary but fixed origin, to an intrinsic system defined by the structure of the mathematical object. Although *numbers* appear to be anchored to a traditional origin at zero, *vectors* incorporate numeric magnitude but are not anchored in space. Creating a special location for nothing by labeling that location "0" is also a violation of the Theory of Relativity. In the universe, there is no privileged perspective. Substituting 0 for nothing is yet another example of human arrogance.

13. **renders that container no longer empty:** Similarly in set theory, ϕ is the name of the empty set and it is explicitly represented by { }. { ϕ } however is not ϕ.

14. **until we make these choices:** P. Lockhart (2012) *Measurement* p.206. This is, of course, a statement of the Participation Principle.

15. **we pull back into a familiar one:** K. Devlin (2006) The Useful and Reliable Illusion of Reality in Mathematics. *Toward a New Epistemology of Mathematics Workshop*, GAP6 Conference 9/14-16/2006.

16. **the operation of distinction defines it and makes it possible:** H. Maturana & F. Varela (1987) *The Tree of Knowledge: The biological roots of human understanding*. In a series of books Varela and Maturana apply distinction-based thinking to biological systems.

17. **In an empty environment we see nothing:** The relationship between object and environment is perceptual, but it is also conceptual. In physical reality there is no emptiness, our environment is full of stuff. When entering an empty virtual environment (via immersive goggles), it is clear that there is nothing. Humanity's ability to see our physical environment as an object, from outside, had to wait until the moon landing in 1969. In 1989, when immersive virtual reality first became popular, humanity had its first direct encounter with an empty environment.

18. **It is this making of distinctions that leads to the world of number:** R. Rucker (1987) *Mind Tools* p.19.

19. **while it changes our interpretation from process to object, from transform to form:** We are in subtle territory here because the processes involved with reading and interpreting textual form have traditionally been ignored within the text itself. We have transgressed from the text to the subtext, from the form to the substance. In contrast, boundary math remains acutely aware of the process of reading itself. *Boundary forms include the reader*. Whether or not we view ourselves as on the outside or as on the inside is a matter of choice. This is just another consequence of cleaving *void*. The **Participation Principle** is an unavoidable consequence of postsymbolic mathematics.

20. **depending upon the focus of the reader**: Explicit outermost boundaries explicitly locate the reader on the outside. Implicit outermost boundaries invite participation on the inside.

On the page we actually have a series of nested boundaries or distinctions, each with a conventional meaning. The page itself distinguishes the book from its content. The page margin distinguishes typographical from non-typographical space. The boundary in Figure 7-1 distinguishes text from display. Each row in the figure distinguishes a numeral. The shell-bracket distinguishes formal from informal text. Each of the round forms distinguishes a whole from nothing. Oh, and yes, this endnote is a clever mechanism to distinguish primary from incidental text, effectively extending the boundaries of the page containing the endnote marker into the endnote section. Had we been using footnotes rather than endnotes, the boundary of the incidental text would have been carved out of the original page.

21. **as it did for millenia, via the accumulation of units:** The addition of one and one is not necessarily obvious. In the *Phaedo*, Plato has Socrates declare:

> I cannot satisfy myself that when one is added to one, the one to which the addition is made becomes two, or that the two units added together make two by reason of the addition. For I cannot understand how, when separated from the other, each of them was one and not two, and now, when they are brought together, the mere juxtaposition of them can be the cause of their becoming two....

22. **interpreted as the base-free logarithm of its contents:** One convenient way to think about logarithms is that the logarithm of a number indicates the number of digits in that number.

23. **then any base is as good as any other:** Lockhart, p.387.

24. **during the comprehensive social integration of number, over which it formerly had a monopoly?**: A. Badiou (2008) *Number and Numbers* §0.10.

25. **of the absence of any grounding by the One:** Badiou, §1.16

Chapter 7

Chapter 8

Transformation

*My aim at that time was to reduce everything
to the smallest possible number
of the simplest possible logical laws.*[1]
— *Gottlob Frege (1884)*

In this chapter we'll explore forms composed of both round and square boundaries. Round boundaries provide a numeric unit and a model of *exponentiation*. Square boundaries provide a non-numeric unit (that we will interpret as negative infinity) and a model of the *logarithm*. Combining round and square boundaries provides a model of multiplication. Due to Unit Accumulation, forms composed only of round boundaries do not reduce. They accumulate. When both types of boundary are present in the same form, a structural pattern rule might apply to reduce the complexity of the form.

Permission to rearrange structure is provided by *pattern axioms* that define allowable transformations. James axioms assure that void-equivalent forms do not change into stable forms, and that stable forms do not disappear into *void*. Rearranged forms maintain the same intended meaning and the same available interpretations.[2] In this chapter we'll be exploring two axioms, **Inversion** and **Arrangement**.

axioms maintain difference

Chapter 8

8.1 Inversion

The inversion of the round boundary ( ) is the square boundary [ ]. The square and round boundaries are co-defined to mutually annihilate, regardless of which contains which. This containment relation is expressed by the Void Inversion axiom.

void inversion

$$([\,]) = [(\,)] = \textit{void}$$

More generally, the Inversion axiom is

inversion
enfold ⇌ *clarify*

$$([A]) = [(A)] = A$$

Void Inversion is thus an application of Inversion for which A = *void*. Inversion is the algebraic form, while Void Inversion is its grounding within the arithmetic of forms.

It will be handy to have a name for each direction of the Inversion axiom. We will call deletion of structure (left to right) `clarify`, and creation of structure (right to left) `enfold`.

We do not yet need to make a distinction between ([ ]) and [( )]. From the perspective of James arithmetic, these nested brackets do not have an intrinsic order of nesting. The visual appearance of a nesting order is another example of an unintended *artifact* introduced by a choice of notation.[3] The artifact predisposes us to think that Void Inversion is two axioms when from the boundary perspective it is one. Sequential order is an artifact of string notation; analogously inside/outside is an artifact of a 2D diagrammatic notation. The plane of the page enforces our outside-of-the-image bias. Imagine standing between the two boundaries. They are clearly different since we must turn to face one or the other. Without the clue of *curvature*, we cannot know whether we are inside or outside. Just like ( )[ ] differs from [ ]( ) when we restrict our perspective to a line, the difference between ([ ]) and [( )] also arises from a restricted perspective, that of presuming we are on the outside of the form.

inside or outside?

Double-Boundaries

We have constructed two types of boundaries that are void-equivalent whenever they form a double-boundary. A **double-boundary** is a structure in which one boundary is the sole contents of another boundary form. A **simply nested form** incorporates no container with more than one content form.

double boundaries
(( ))
([ ])
[( )]
[[ ]]

The round/square pair of boundaries are jointly called an **inversion pair** when there are no intervening forms. Inversion pairs are void-equivalent and can be freely deleted. They are structural noise within an intentional signal. Due to Inversion, the only simply nested forms that do not reduce consist of purely either round- or square-brackets. Simply nested forms will take on greater structural variety after we introduce the angle-bracket in Chapter 10.

stable simply nested forms
((( )))
[[[[[ ]]]]]

Inversion boundaries are void-equivalent only when they are simply nested. Should the outer boundary contain other contents, the pair of boundaries do not reduce.

([A]) = A

(B [A]) *does not reduce*

Round/square boundary pairs that do not reduce, such as (A [B]), are a **central organizing concept** in James algebra. These forms are called *frames*, and will be explored a little later in the chapter.

Inversion provides an assured outermost boundary for any generic collection of boundary forms.

A B C = ([A B C]) = [(A B C)]

Should we chose, now we can abandon the ( ) notation, since any collection of content can be expressed as having an outermost round or an outermost square boundary.

Void-equivalent Forms

Since James transformations are void-based, some specific configurations of containers may transform by vanishing. A variable might therefore stand in place of a form that has no meaning. This is a delicate distinction, between arbitrary structures that at some point become stable after reduction, and void-equivalent structures that are essentially meaningless since they reduce to *void*. The transformation patterns of a particular void-based boundary system partition constructible forms into those that have meaning and those that do not have meaning by distinguishing forms that are grounded in each category: existent and void-equivalent.

meaningful: ( )
meaningless: [( )]

Another way of saying this is that within a void-based system, some structural forms are *illusions*. For example, the form ([ ]) is void-equivalent under the Inversion axiom. Upon reduction, ([ ]) ceases to exist. This does not mean that ([ ]) has no value within an interpretation. The partial expression "+ 0", for example, is void-equivalent in conventional arithmetic because it does not interact with value. What the void-based approach guarantees is that whatever consistent interpretation is placed upon a void-equivalent form, that interpretation will not interact with the value of stable forms.

8.2 Pattern Variables

We have introduced the Roman letter A as a *pattern variable*. Variables stand in place of arbitrary expressions, and thus move notation from an arithmetic to an algebra. For our purposes, Roman letters, both large and small, stand in place of **arbitrary patterns of containment**. A pattern-variable applies to all forms, all permitted structures within a pattern algebra, including patterns that may have a non-numeric interpretation and including the non-pattern *void*. James forms are both object and operator, broadening the concept of a variable not only

Transformation

to stand in place of objects that can be reduced, but also to stand in place of the operations that achieve transformation. In conventional arithmetic, an expression such as z + 3 uses the variable z as an arbitrary object, a form that represents a number. There is no concept of an operator variable (until more advanced math topics such as second-order logic). For example, in

$$x ⁂ y = y ⁂ x$$

the symbol ⁂ could be taken to assert commutativity of both + and x. In contrast, a pattern variable does not need to distinguish object or operator, it matches structure, not function, not value and not meaning.

Since all forms have an outermost boundary, each pattern of containment is bounded by a single container. There is no constraint on the numeric value of a form because numeric value is not an intrinsic property (it's an interpretation). James algebra includes non-numeric container forms as well, so that the language of James forms is in some ways broader than the language of numbers.

The *context* of a variable limits and defines the values that it can be assigned. The algebraic approach usually presumes that variables can be any value unless constrained by their context. The central tool that applies constraint is the equal sign, =. James equality is defined by permissible transformation patterns which in turn are specified by the axioms. James variables then are **algebraic pattern variables** for which the specific form of their containment limits their available interpretations.

The conventional approach to transformation over sets of objects is to identify a domain to which all particular expressions belong. A variable possesses **universal quantification** when it can stand in place of any member of that domain. The domain of James forms is finite and physically constructible configurations of containers.[4] This domain is more expressive than that of the rational

void ☞
((([ ])[[ ]]))

1/2 ☞ (<[2]>)

√2 ☞
(([[2]]<[2]>))

∞/∞ ☞
([<[ ]>]<[<[ ]>]>)

numbers since James forms also represent real numbers such as π, √n, log n, and sin θ; non-numeric concepts such as infinity; operations such as multiply; and illusions that are void-equivalent. With such inclusions, it is better just to abandon the more refined logical concepts of *domain* and *quantification*, since they amount essentially to a declaration of Existence.

James variables are more like labels. Labels do not act like abstract universally quantified variables. For example, when we forget a person's name we have forgotten only a label, not the person. Losing the binding of a conventional mathematical variable loses both the name and the referent. When a variable serves as a label, we say that the variable has been **instantiated** or **bound**. A label without a referent is **free**. The subtlety again is that a variable cannot be bound to *void*, so we must be careful about whether a variable stands in place of a stable but arbitrary structure or in place of an illusion. In any event, the variable identifies structure but not meaning.

quantification becomes qualification

With no axioms every form has meaning. When a void-based transformation is taken as an axiom, the domain becomes permeated with vacuous structures that shrink the domain itself. This means that **existential quantification**, the assertion that a variable stands in place of at least one stable form, depends upon transformation rules as part of its definition. We are no longer free to conceptually distinguish variables from axioms.

When a system includes void-based transformation patterns, computation and proof proceed via **deletion of irrelevant structure**. Simplification separates meaning from structural junk. This is one way that void-based arithmetic clarifies the accretion of rules and structural definitions incorporated into conventional arithmetic. There is never any utility in maintaining junk once it is identified. In contrast, the symbolic approach presumes that all forms can be represented and that all manipulations are meaningful. For example,

$$([\,]) = \textit{void} \quad \mathbb{\tiny\mathcal{F}} \quad \#^{\log\#0} = 0 \qquad \#^{\log\# x} = x$$

Symbolically, the right-hand equation is a legitimate expression that is equal to zero by virtue of a specific structural rule defined for exponential expressions. The left-hand equation gives permission to discard a meaningless form.

Interpretation

Inversion pairs can be interpreted as any pair of conventional inverse functions, for example, ( ) as double and [ ] as half.

$$([\ ]) \quad \mathbb{\tiny\mathcal{F}} \quad \text{double(half(zero))} = \text{zero}$$
$$([4]) \quad \mathbb{\tiny\mathcal{F}} \quad \text{double(half(four))} = \text{four}$$
$$[(A)] \quad \mathbb{\tiny\mathcal{F}} \quad \text{half(double(A))} \ \ = \ A$$

Alternatively, ( ) could be interpreted as the successor of a natural number, while [ ] as the predecessor. We can also choose to engage in a non-numeric interpretation such as ( ) standing in place of "open the door" and [ ] representing "close the door". Each unit boundary defines a new state of the door, while void inversion returns the door to its original state.[5]

$$([N]) = N$$
$$\mathbb{\tiny\mathcal{F}}$$
$$+1(-1(N)) = N$$

Alternating depths of containment can be mapped onto even/odd, affirm/negate, FORALL/THEREEXISTS, AND/OR, and other formal duals. Without interpretation, we simply have two apparently different types of boundary that eliminate one another by containment.

Chapter 8

Alternating Contexts

In our chosen interpretation of exponential and logarithm, round- and square-brackets have an associated base that they implicitly share. The round-bracket ( ) *contains* exponents. The variable A in (A) is interpreted as an exponent while the round-bracket represents the arbitrary base, $\#^A$. Inversely, the square-bracket [ ] contains logarithmic content. The variable A in [A] is the form we are identifying the logarithm of, while the square-bracket represents the logarithmic function, $\log_\# A$.

The separate contents of round-brackets can be seen to represent multiplications. For example,

$$(A\ B\ C) \quad \text{☞} \quad \#^{A+B+C} = \#^A \times \#^B \times \#^C$$

The form of multiplication is more clearly illustrated when the variables are themselves within a square-bracket.

$$([A][B][C]) \quad \text{☞} \quad \#^{\log_\# A + \log_\# B + \log_\# C}$$
$$= \#^{\log_\# A} \times \#^{\log_\# B} \times \#^{\log_\# C} = A \times B \times C$$

Every form can be interpreted as a multiplication. In the case of adding three forms, each of A, B, and C can be enclosed with an application of enfold, and if necessary the entire form can also be enfolded. Here we convert addition into multiplication by applying enfold four times:

$$A\ B\ C = [(\ [(A)]\ [(B)]\ [(C)]\ )]$$
$$\text{☞} \quad A + B + C = \log_\# (\#^A \times \#^B \times \#^C)$$

Figure 8-1 shows that many different conventional representations can be seen as a single boundary form embellished with inversion pairs at specific locations.

When square-brackets are interpreted as $\log_\#$, we might say that ( ) is 1, and that the logarithm of 1 in any base is 0.

$$[(\ )] = \textit{void} \quad \text{☞} \quad \log_\# 1 = 0$$
$$([\ ]) = \textit{void} \quad \text{☞} \quad \#^{\log_\# 0} = 0$$

form	☞	expression of 2×3
([oo][ooo])		$\#^{\log_\# 2\, +\, \log_\# 3}$
([([oo])][ooo])		$\#^{\log_\# 2} \times 3$
([oo][([ooo])])		$2 \times \#^{\log_\# 3}$
([([oo])][([ooo])])		$\#^{\log_\# 2} \times \#^{\log_\# 3}$
[(([[(oo)]][ooo]))]		$\log_\# (\#^2)^3$

Figure 8-1: *Varieties of* 2×3

In the general case,

[(A)] = A ☞ $\log_\# \#^A = A \times \log_\# \# = A \times 1 = A$

([A]) = A ☞ $\#^{\log_\# A} = A$

To maintain consistency, the interpretation of the empty square-bracket is somewhat surprising.

[ ] ☞ $\log_\# 0 = -\infty$

This interpretation means that, from a conventional perspective, James algebra incorporates negative infinity as a basic type of unit, providing computation with infinities at the core of the algebra.

([ ]) ☞ $\#^{\log_\# 0} = \#^{-\infty} = 1/\#^{\infty} = 0$

Like *void*, the concept of infinity is also *non-numeric*. Structurally, James algebra freely mixes numeric and non-numeric forms, highlighting again that it is the interpretation that makes James algebra about numbers.

The separate contents of square-brackets do not match the pattern of the Inversion axiom and thus do not transform. The corresponding logarithm does not act separately upon expressions that are added together. For example,

[A B C] ☞ $\log_\# (A + B + C)$

Chapter 8

Logarithmic Foundation

What does it mean to define brackets as logarithmic and exponential operators? The perspective we are taking is that *exponentials and sums are equally fundamental*, while multiplication is a composite and therefore derivative operation.

John Napier
1550–1617

Historically, logarithms arrived late on the scene, an invention of Napier in the early seventeenth century. Napier's new function mapped multiplication onto addition, and thus showed a way that multiplication could be converted into addition. The original use of logarithms was to simplify the computation of products of trigonometric functions that occurred within planetary astronomy. Napier incidentally introduced the wide spread use of approximate decimal fractions as a convenient way to record and publish large look-up tables of logarithmic values.[6]

(N) ☞
$\#^N = \# \times .._N.. \times \#$

When ( ) is viewed as an operator on its contents, a conventional interpretation of exponentiation would suggest that (A) defines A replicas of an arbitrary number (the base) all multiplied together. A difficulty with this interpretation is that A is not constrained to be a positive integer. If A were void-equivalent, for example, what would it mean to multiply a number by itself no times? If A were a transcendental, say π, what does it mean to generate π replicas of some number? Even if A were something simple, like the fraction ½, what would it mean to generate half a replica of a number? If we take the conventional interpretation that $\#^2 = \# \times \#$, how do we explain that # identifies *two* numbers, ± #. Finally when the arbitrary base is negative it becomes necessary to introduce the fundamentally new domain of imaginary numbers.[7]

( )

(π)

(½)

(2)

A subtle constraint has been introduced by the interpretation of [ ]: *# is greater than* 1. When # is between 0 and 1, say ½, the fractional base changes the sign of the result.

$\log_\# 0 = -\infty$ $\qquad\qquad \log_{1/\#} 0 = +\infty$

For now we'll assume that # > 1 when # is the base of a logarithm. Notice that when # is used for the base of an exponential, there is a different restriction on the interpretation, that # > 0.

Changes in the arbitrary base itself can be better appreciated from the perspective of the square-bracket. The empty square-bracket provides the "point at infinity" similar to the point-at-infinity that is fundamental to projective geometry and to homogeneous matrices. Logarithms are inherently transcendental, beyond the scope of algebraic operations, and certainly incommensurate with whole numbers. Again the same questions about the arbitrary base arise, but this time outside of the idea of the unit One, outside of counting, and outside of the Additive Principle. Logarithms convert the curious questions about void and fractional and transcendental *powers* into curious questions about void and unitary and negative and transcendental *bases*. Exploring exotic bases has been postponed until Volume III.

8.3 Arrangement

To complete an operational description of round- and square-brackets, we will need an axiom that specifies how the two boundaries interact in configurations more complex than the double nesting of Inversion. Here is the James axiom of Arrangement. The axiom specifies an invariant structure across frames.

$$(A \ [B \ C]) = (A \ [B]) \ (A \ [C])$$

arrangement
collect ⇄ *disperse*

The simplifying direction (from right to left) is collect, and the expanding direction (from left to right) is disperse.

The Arrangement axiom gives permission to take multiple forms out of the interior content of a frame and put them into separate frames, so long as we carry their

Chapter 8

context — i.e. their frame type, (A [...]) — along with them. Arrangement is roughly equivalent to conventional distribution, with the difference that the conventional distributive rule distributes multiplication over addition,

$$A \times (B + C) = (A \times B) + (A \times C)$$

while our interpretation of Arrangement distributes addition of logarithms over exponents. Due to the awkward conventional notation for exponents, the interpretation of Arrangement in standard notation (plus the base-free #) is not pretty:

$$(A\ [B\ C]) = (A\ [B])\ (A\ [C]) \quad ☞$$
$$\#^{(A + \log_\# (B + C))} = \#^{(A + \log_\# B)} + \#^{(A + \log_\# C)}$$

Consider what happens when the context A is *void*:

$$(\ [B\ C]) = (\ [B])\ (\ [C])$$

a frame
(*type* [*contents*])

In this case, Arrangement turns into three applications of enfold.[8] When A is not void-equivalent, the structure of Arrangement is very much driven by our current interpretation. It is this pattern that provides the **only structural complexity** within the James form. It is the only pattern that permits *rearrangement* as opposed to deletion/creation. It is the only pattern that differentiates the content of James frames. And it is the only pattern that introduces replication (of the frame-type A) without changing cardinality.

Although an equality, the different transformation directions of Arrangement have completely different objectives. Collect gathers forms that share the same type of frame, in effect making the frame-type a global property of the content forms. Disperse separates the frame-content so that they can be transformed separately as individual frames.

Generality

Arrangement is much broader than conventional distribution since container forms support multiple conventional interpretations. Consider the simplest case of distribution of multiplication over addition:

$$1 \times (1 + 1) = (1 \times 1) + (1 \times 1)$$

Converting this into iconic notation, we get

$$([o][oo]) = ([o][o]) ([o][o])$$

o is 1. [o] is 0. Successive applications of Inversion provide a comforting result,

$$([o][oo]) = ([o][o]) ([o][o]) \qquad \textit{disperse}$$
$$([oo]) = () () \qquad \textit{clarify}$$
$$oo = o o \qquad \textit{clarify}$$

Alternatively, we could express the distributive rule in a generic rather than a multiplicative form:

$$([A][B\ C]) = ([A][B]) ([A][C]) \qquad \textit{multiplicative}$$
$$(\ A\ [B\ C]) = (\ A\ [B]) (\ A\ [C]) \qquad \textit{generic}$$

This is not a significant structural difference, since any form can be bounded by [ ] via the Inversion axiom: A = [(A)].

There is a form of Arrangement that looks like conventional factoring, so we could call it the factored interpretation.

$$(A\ [(B)(C)]) = (A\ B) (A\ C) \qquad \textit{factored}$$

This form is decidedly not factoring, the resemblance is purely superficial. Specifically, [(B)(C)] is not a multiplication, it is a logarithm.

$$[(B)(C)] \quad \mathrel{\mathord{\rightarrow}} \quad \log_{\#} (\#^B + \#^C)$$

Log does not distribute over addition, so this conventional form is as simple as it gets.[9]

Chapter 8

The factored version of Arrangement can be cast into the form of multiplication, yielding a conventional but unusual rule of exponents.

factored and multiplicative

$$([(A)][(B)(C)]) = (A\ B)\ (A\ C)$$
☞ $\quad \#^A \times (\#^B + \#^C) = \#^{A+B} + \#^{A+C}$

To make this interpretation more familiar to the eye, we can partition it over two conventional rules: distribution followed by the product rule of exponents.

$$N^A\ (N^B + N^C) = N^A N^B + N^A N^C = N^{A+B} + N^{A+C}$$

Arrangement is limited to the contents of round-brackets, and to the structure of a James frame. The nesting of boundaries below is reversed, with round enclosed by square.

$$[A\ (B\ C)] =?= [A\ (B)]\ [A\ (C)]$$

We can analyze this potential equation by looking at the same form in the arithmetic. We'll render B and C *void*, and take A as a single round unit.

$$[(\ )(\ )] =?= [(\ )(\ )]\ [(\ )(\ )]$$

Is [( )( )] blind to replication? The Accumulation axiom does not permit the form ( )( ) to condense, so [( )( )] cannot be void-equivalent. Only non-numeric forms are insensitive to numeric replication.

Conventional explanations declare that multiplication distributes over addition, but addition does not distribute over multiplication. This might make sense, however it is not an *explanation*. The reason we cannot distribute addition over multiplication has little to do with the operations of addition or multiplication. A more fundamental reason is that construction of replicas of units changes the cardinality of the collection of units (i.e. it changes *value*). Distribution of multiplication over addition, in contrast, replicates frame-types and does not impact magnitude.[10]

Parallel Arrangement

In general, there is no constraint on how many forms B,C,... can be collected or dispersed at the same time. The general form of Arrangement is

([A][B...Z]) = ([A][B])...([A][Z])   **parallel arrangement**

The ellipsis, ..., is finite, there is no intention to introduce infinite collections. However, at least with regard to countable infinities, Whitehead and Russell say that

> Cardinal arithmetic is usually conceived in connection with finite numbers, but its general laws hold equally for infinite numbers, and are most easily proved without any mention of the distinction between finite and infinite.[11]

We are following in Russell and Whitehead's footsteps by developing a calculus of pure pattern and form that connects to numeric concepts only through interpretation.

Arrangement is the primary source of unavoidable complexity in both conventional and boundary calculi. The replication of the form A in Arrangement has both computational and representational implications.[12] The **Principle of Arrangement** makes this explicit.

> **Arrangement:** *Arrangement is the sole source of complexity in elementary mathematics.*

The other James axioms identify void-equivalent forms that can contribute clutter but not computational complexity. We will see in Chapter 9 that Arrangement is the primary mechanism that differentiates conventional addition from multiplication. More fundamentally, Arrangement is the **organizing principle** that permits our interpretation to unite the common laws of arithmetic. There is a certain degeneracy in having only one axiom that mentions more than one variable, since that axiom *must* account for the entire range of compound expressions in an arithmetic.[13]

8.4 Frames

simply nested
((( )))

accumulating
( )( )( )

Frames are the structural framework underlying both Inversion and Arrangement, indeed underlying most stable James forms. Frames are forms for which an outer round-bracket contains at least two content forms, at least one of which is enclosed in a square-bracket. For example (A [B]). Otherwise, a numeric form is either simply nested or accumulating.

A B C =
[(A [(B C)])]

Frames provide a conceptual scaffolding as well as a structural template. The axioms of James algebra act on the round and square frame boundaries together, not on each boundary separately. This permits the base-free interpretation of exponents and logarithms. When boundary configurations are interpreted as numeric operations, the conventional axioms of algebra can be organized as varieties of frames.

The **generic structure** of a frame is

generic frame
(*frame-type* [*frame-content*])

The **frame-type** is structure *between* round- and square-brackets. The generic template collapses if either the type or the content is void-equivalent. There are categories of frames that all have the same frame-type. In Arrangement collect gathers together *framed-content* that has in common the same *frame-type*.

arrangement
(A [B C]) = (A [B]) (A [C])

The primary impact of disperse, although it does disperse framed content, is to replicate the *frame-type*. Frames are thus forms that support structural replication without impacting numeric value.

Frame Types

Figure 8-2 identifies several types of frames, most of which we will see in the following chapters.

Transformation

([A])	= A	**inversion frame**
(A [ ])	= *void*	**dominion frame**
(A [B C])	= (A [B]) (A [C])	**arrangement frame**
(o [B])		**unit magnitude frame**
(<o>[B])		**decimal frame**
(A [B])		**magnitude frame**
([A][o])		**indication frame**
([A][N])		**cardinality frame**

Figure 8-2: *Types of frames*

Since our interpretation takes round-brackets to be powers and square-brackets to be logarithms, a generic frame represents an exponential expression, $B\#^A$.

$$(\ A \ \ [B]) \ ☞ \ \#^{A+\log_\# B} = \#^A \times B$$

$$([(A)][B]) \ ☞ \ \#^A \times B$$

The *cardinality* frame expresses multiplication in general.

$$([form][cardinality]) \ ☞ \ form \times cardinality$$

cardinality frame

The cardinality frame disperses replicas of generic forms.

$$([A][N]) = A \ ..\!_N.. \ A$$

The Indication frame is the base-case of cardinality, and in Chapter 9 plays a central role in counting.

$$([A][o]) \quad ☞ \quad A \times 1 = A$$

The *magnitude* frame expresses the exponential structure of polynomial digits.

$$(power \ [value]) \ ☞ \ value \times base^{power}$$

magnitude frame

The magnitude frame thus resembles scientific notation, and disperses replicas of exponential forms.

$$(\ A \ [N]) = (A)..\!_N..(A)$$

Chapter 8

Unitary magnitude multiplies a given magnitude A by the elected base #.

$$(\circ [A]) \quad \mathrel{\text{☞}} \quad \#^{1+\log_\# A} = \# \times A$$

Here's an example of the general form of an exponential with an explicit base.

clarify
$$A \times B^N \quad \mathrel{\text{☞}} \quad ([A][(([B])[N]))])$$
$$([A] \ (([B])[N]) \ )$$

Embedding the base B into the boundary forms yields the magnitude frame. Recall the form of the base: B = (( ))

implicit base-B
base ☞ (( ))

$$([A](([(( ))])[N])) = ([A] \ N)$$

Here's another example, this one rather extreme.[14β] I'll omit the base-# reminder for the logarithms.

disperse
clarify
$$A^{\#^B + \#^C} \quad \mathrel{\text{☞}} \quad ((([A]] \ [(B)(C)]))$$
$$((([A]][(B)]) \ ([[A]][(C)]))$$
$$((([A]] \ B \ ) ([[A]] \ C \ ))$$
$$\mathrel{\text{☞}} \quad \#^{\#^{(\log\log A)+B} + \#^{(\log\log A)+C}}$$

For comparison, a direct reading of the James form of the example as exponents and logarithms is

$$A^{\#^B + \#^C} \quad \mathrel{\text{☞}} \quad ((([A]] \ [(B)(C)]))$$
$$\mathrel{\text{☞}} \quad \#^{\#^{\log\log A + \log(\#^B + \#^C)}}$$

Here Arrangement is constructing, at the exponent level, the multiplication of $\log_\# A$ by $\#^B$, and adding it to the multiplication of $\log_\# A$ by $\#^C$. Yes, this form would rarely be useful. The point is not that James Arrangement can handle weird exponential forms, rather that it suggests new forms of counting, forms that do not end up interpretable as multiplications. It is, after all, our conventional notation that generates the difficult to read expressions. And a clumsy notation will suppress exploration of some lines of thought. Bertrand Russell:

A good notation has a subtlety and suggestiveness which at times make it almost seem like a live teacher.[15]

8.5 Remarks

Recall that the James forms are not designed to be as efficient for reading as our conventional notations that have evolved over thousands of years. Conventional notation is relatively easy to read and difficult to work with. James forms exchange readability for convenience. Our grander objective is to explore a simpler model for thinking about mathematics and reality. Sometimes this requires expanding a highly condensed notation in order to expose its basic structure.

James algebra is thus far constructed from two structural pattern transformations: **Inversion** and **Arrangement**. Arrangement handles all the work of computation. Inversion both cleans up the results of Arrangement and serves to mediate the variety of interpretations that derive from a fully reduced James form. In Chapter 10 we will add a third and final pattern to incorporate inverse operations.

The diversity of concepts that have evolved within conventional arithmetic are each, from the James perspective, the same concept in different contexts. Apparently different operations (+, ×, ^) are unified under the common concept of a *frame*. The interplay between the two boundaries, ( ) and [ ], provides both theoretical and computational definitions. The distinction between a mathematical object, such as the integer 2, and an operation, such as adding 2, is eliminated. We can interpret the ground object ( ) as raising any base to the zero power, or as a numeric unit o.

$$( ) = o \quad \mathbb{\mathbb{F}} \quad \#^0 = 1$$

The interaction of round and square containers defined by Inversion and Arrangement provides sufficient mechanism to construct and describe the behavior of whole numbers.

Chapter 8

We can interpret the ground object [ ], *if we must*, as the arbitrary-base logarithm of zero, or as a non-numeric unit.

$$[\ ] \quad ☞ \quad \log_\# 0 = -\infty$$

A possible *goal* for exploring this exotic algebra is to learn something new (and unexpected) about how to think about numbers. The *motivation* is a suspicion that our current notation for numbers, having accreted over thousands of years, may be harboring both computational inefficiency and conceptual obfuscation. In particular, do the Rules of Algebra actually define the essence of numbers? We have taken this path driven by two observations. First, arithmetic and algebra seem to be very difficult for school children to comprehend. Perhaps their native intelligence is telling us something about the quality of our abstract numeric concepts. Second, both ancient and indigenous number systems are based on the Additive Principle. Are we entirely justified in abandoning the biological and sociological origins of arithmetic? Is it wise to replace native arithmetic with an abstract sequential symbol system that has only recently been popularized and is acknowledged to harbor conceptual difficulties (e.g. the axioms of set theory, the paradoxes, Gödel incompleteness, the reliance on non-numeric and non-existent infinities, absence of parallelism, a deceptive symbolic 0, etc.)?

It is true that growing a technical field from original concept to extreme abstraction can be advantageous. But not so for a societal tool backed by 8000 years of cultural evolution. And not a tool that is required content for seven-year-old children to learn. And not a tool that most elementary and high school *teachers* despise.

The objective is not to overturn conventional knowledge of numbers, nor to suggest that the world-wide perspective on numeric computation is inadequate. However, human knowledge has long been known to be, well, laden with confusion, usually brought on by inappropriate arrogance driven by particularly narrow vision. An invariant of human knowledge is that what was universally known to be certain a century prior, might be known to be folly in the centuries that follow.[16]

Was the mathematics that co-evolved with commerce a folly, and extreme abstraction its correction? Or is our latest excursion into non-humane mathematical abstraction what the future will see as folly? One thing is clear: there is absolutely no empirical evidence that knowledge of high-school symbolic algebra is useful for survival or success or clearer thinking. It is also known that about one in twenty adults in our society will ever use or need or benefit from the current high school math curriculum.

We next revisit the historical origin of arithmetic, returning to the concept of **counting as tallying units**, and to tallying as collecting in a container. Prior to being able to count, we will need to be able to construct identical replicas that collected together provide the concept of cardinality. As is usually the case with mathematics, we start our four-year-olds in the middle of the story of numbers, disguising the beginning, as every two year-old knows, by pointing at something. Adding, multiplying and exponentiation arise naturally as recursive pointing at the act of pointing.

Chapter 8

Endnotes

1. **opening quote:** G. Frege (1884) *The Concept of Number* §91.

2. **maintain the same intended meaning and the same available interpretations:** The axioms of James algebra were designed to support an interpretation as numeric forms, although they also support other interpretations. The design choices themselves were grounded not in numerics but in Spencer Brown's boundary logic, the *Laws of Form*. Chapter 15 provides a map between James algebra and boundary logic.

3. **an unintended artifact introduced by a choice of notation:** Our cultural perspective of inside/outside is so dominant that we have no textual or planar representation available to express the idea that *the side of a boundary we view from is a choice*. Such is the bias of our belief that the world is outside of us and we are inside the world. It took going to the Moon to show humanity that the environment we are inside can also be viewed from outside. It takes introspection to understand that the interior mind and the exterior body reflect different perspectives on the same boundary.

4. **finite and physically constructible configurations of containers:** Pure physicality must be augmented by multiple reference in the form of replicas of labels or pointers to a concrete object. The mathematical structure of containers is discussed in Volume II and the particular properties of replicas in Chapter 9.

5. **void inversion returns the door to its original state:** Of course, the algebra of opening and closing doors is not rich. However the same structure occurs in logic, as double negation. A second door provides a choice of which door to open, a physical structure equivalent to logical OR. The expressibility and formal structure of propositional calculus can be fully modeled by the ability to choose one of two doors not to open or close (logical NOR).

6. **publish large look-up tables of logarithmic values:** Polynomials in the form of quadratics show up very early in the history of algebra. Perhaps because polynomials express the relationship between the perimeter of rectangle and its area, techniques for solving quadratic equations appear in Babylonian mathematics 4000 years ago. Competitions for solving higher order polynomials fascinated European academics over 700 years ago. The cultural evolution of algebra largely ignored logarithmic forms until the sixteenth century.

Transformation

7. **led to the fundamentally new domain of imaginary numbers**: As described in Chapter 2 and further examined in Volume III, the design choice that $-\times-=-$ is consistent when multiplication is taken to be non-commutative,

$$a \times b = - b \times a$$

The change makes it not possible to express what we call imaginary numbers.

$$i = \sqrt{-1} = -1 \quad \textit{since} \quad -1 \times -1 = -1$$

When $-\times-=-$, the entire domain of imaginaries fails to be brought into representational existence and is replaced by a different kind of imagining.

8. **Arrangement turns into three applications of enfold**: You may have noticed that some varieties of Arrangement combine Inversion with the generic form of Arrangement. I've alluded to a subtle difference, that we can view boundaries from either side. The occurrences of Inversion within Arrangement can also be read as a change only in viewing perspective, alternating between looking at a form from the outside and looking at a form from the inside. This reading skill is better explored from boundary logic, without the complexity of forms that accumulate.

9. **this conventional form is as simple as it gets**: Log does distribute across multiplication, a fundamental feature that allows logarithms to convert multiplication into addition. Addition itself is the distribution of logarithms across multiplied exponents.

10. **in contrast, replicates frame-types and does not impact magnitude**: Frame-types are analogous to the base of a conventional number, while framed content is analogous to magnitude.

11. **without any mention of the distinction between finite and infinite**: A. Whitehead & B. Russell (1910) *Principia Mathematica* Preface p.vi.

12. **Arrangement has both computational and representational implications**: This is also the case in logic. Distribution of OR over AND is the primary transformation for converting logic expressions into their canonical two-level nested form, called **conjugate normal form**. Distributive replication grows the size of the logical expression exponentially. Interestingly, there is no canonical form for deeply nested logical expressions. Nested within our mathematical culture is an aversion to nested logical deduction.

Chapter 8

13. **that axiom must account for entire range of compound expressions in an arithmetic:** In his 1994 Master's thesis from Maharishi University *Natural Numbers and Finite Sets Derived from G. Spencer Brown's Laws of Form*, Jack Engstrom demonstrates the dependence of both natural numbers and sets upon Spencer Brown's concept of distribution, what we are calling Arrangement.

14. **Here's another example, this one rather extreme:** Yep, I made this one up and it looks weird in conventional notation.

15. **at times make it almost seem like a live teacher:** B. Russell, Introduction to L. Wittgenstein (1922) *Tractatus Logico-Philosophicus* p.15.

16. **will be known to be folly in the centuries that follow:** Leonard Kronecker (circa 1880) espoused the radical viewpoint that mathematics is an invention of man (!). He believed that only natural numbers were divine. The majority of mathematicians at the time believed firmly that *all* mathematics is the work of God, aka the Platonic reality. From our perspective, Kronecker did not go far enough. Even the natural numbers are not sacred.

Chapter 9

Accumulation

The most elementary as well as the highest mathematics are economically-ordered experiences of counting, put in forms ready for use.[1]
— *Ernst Mach (1895)*

The enterprise of arithmetic is built upon counting up various accumulations of marks, sums, multiplications and exponentiations. Here's Brian Rotman:

> What seems universally accepted is that numbers are inconceivable — practically, experientially, conceptually, semiotically, historically — in the absence of *counting*....And, as an activity, counting works through — it *is* — significant repetition.[2]

However counting itself is not the basis of arithmetic. Counting is a quite complicated affair consisting of the cognitive acts of identifying and categorizing, followed by the computational acts of indicating, fusing and labeling.

To be able to accumulate N observations, we must necessarily see the items as *replicas sharing some common feature*. A **replica** is a cognitive reconstruction of a physical object ready to be stripped of any or all detail. A collection of replicas is a cognitive **category**. We create

Chapter 9

ensembles by abstracting replicas to fit into a category and then ignoring their differences. The main idea in this chapter is that *we can count only ensembles*. Unique physical objects do not accumulate, they remain unique, even as we construct replicas of them.

To count twelve items of furniture for example, we first identify each item by constructing a cognitive replica. Then we elect to consider our replicas of the table together with six chairs, and the three bookshelves and the two bureaus as being of a single type, they are each an "item of furniture". Collections can be *tallied* only by virtue of a common intrinsic property that they share. Minimally, that communality is being in the same container; in the case of furniture, perhaps being in the same room. When we use a container to impose a countable communality upon its contents, we attribute the property of **cardinality** to that container. Cardinality is therefore an external perspective; individual contents remain as featureless units. Logician Richard Heck:

tallies are identical

> Children, and adults, seem to be able to understand some attributions of cardinality quite independently of any connections with counting.[3]

As an example, children use the rhyme "Eeny, meeny, miny, moe" as a selection device in games, without implying any particular ordering. The rhyme depends exclusively upon one-to-one correspondence, and soon turns into a cadence of words "catch a tiger by the toe". The result of the process identifies one child who, for the purposes of the game, is "it". "It" is not necessarily associated with "nth" or even "last". In the counting-out rhyme "One potato, two potato..." participants are successively eliminated, again with no intention of determining an ordering or a specific cardinality, even though the numerals are named during the selection process.

9.1 Sets and Fusions

Cardinality also refers to the number of members within a set. The primary difference is that sets contain unique members while tallies are identical. Set theory creates collections that can be counted by including a batch of unique things within a single set. Although this idea is taken to be the foundation of natural numbers, it leads to some bizarre consequences that we will look at in Volume II. When numbers were first formalized (i.e. clearly defined) at the end of the nineteenth century, they were thought of as a **fusion** of indistinguishable tally marks. Dedekind and Frege both built numbers from fusion rather than sets.[4] Both tallies and fusions were explored in Chapter 2 on unit-ensembles.

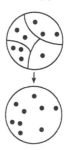

addition by fusion

When sets became philosophically popular about 1910-1920, fusions were abandoned. The motivation, perhaps, was that putting-replicas-together worked well for whole numbers, but not for *everything* in mathematics. Here, fortunately, we are only trying to understand common numbers, so we'll avoid (at least until later) the bewildering consequences of using infinite sets to describe counting, or for that matter, any cognitive act.

The difference between fusions and sets has directly impacted how we have been taught to think about numbers. In set theory, the **axiom of choice** implies that we can put the members of a set in an unambiguous order, a *sequence*. From there we can proceed to associate each object with the sequence of natural numbers, indexing the objects as 1, 2, 3,..... A theme in our current exploration is that sequential thinking and temporal indexing are historical artifacts. The replicas in a container are all there at the same time, concurrently; we do not need to put them into a tidy linear sequence in order to be able to determine cardinality. It is our intellectual history of putting words and ideas in lines that entices us to continue to think that mathematics too must be well-ordered.

Chapter 9

COUNTABLE	NOT COUNTABLE
separate discrete objects	webs of interrelations
frozen pictures	fluid motion
clear boundaries	mixtures
physical discontinuity	cognitive activity
events stopped in time	events changing over time
simple pieces	complex structures

Figure 9-1: *What can be counted?*

A fundamental difference between the concepts of set theory and those of boundary math is that boundary math incorporates cognitive acts while set theory does its best to eliminate the idea that a human is involved in mathematical form. Since boundary math is based upon distinction as its singular foundational concept, and since distinction is a cognitive act, boundary math is deeply connected to the human who is constructing the math. This chapter makes that connection overt by introducing the cognitive concepts of replica and category.

Constructivist mathematicians such as Kronecker and Brouwer believed that the foundation of mathematics rested in our *intuition* of natural numbers. We are digging under that belief to expose a foundation based on a single idea: distinction. Davis and Hurst state the issue clearly:

> The dogma that the intuition of the natural number system is universal is not tenable in the light of historical, pedagogical, or anthropological experience. The natural number system seems an innate intuition only to mathematicians so sophisticated they cannot remember or conceive of the time before they acquired it; and so isolated that they never have to communicate seriously with people (still no doubt the majority of the human race) who have not internalized this set of ideas and made it intuitive.[5]

A final difference between sets and bounded tallies is that set theory is thoroughly dependent upon first-order logic for its definitions and its axioms, whereas boundary forms instead depend upon pattern matching.

9.2 Counting

We are about to explore the conceptual foundations of natural numbers in order to understand that numbers are cognitive constructions, and decidedly not part of the natural world.[6] In reality, everything is intricately woven into **one unified whole**. Reality is not fragmented, nor is it indexed, nor is it linear, nor is it enumerated. It is human cognition that first constructs objects and then abstracts them sufficiently to be able to determine how many. What we perceive as objects are differences that we impose, boundaries that we construct in support of our own unique perspectives. We construct an object by severing it from a woven context.

an object is a severed network

network

object

Very few things actually meet the criterion of being countable. Figure 9-1 provides a listing of some of the things we can and cannot count. Foremost is the idea that we ignore sufficient detail to be able to put things into the same category so that one of them and one of them makes two of the same them. Even the word *object* is misleading, since it presumes that we have already forgotten sufficient detail to be able to detach our personal definition of thing from the web of interrelations to see it as a discrete thing.[7] We have drawn a distinction around selected relations.

Objects are defined by their interactions rather than by their independent properties. Put colorfully, a hand is not five fingers, it is thirty-two relations between fingers. The uncountable constructions of our minds illustrate clearly the limitations of arithmetic. We cannot count thoughts, feelings, beliefs, the wind, a river, sunshine, frost, the internet, a conversation or the web of life. We can count peaks but not mountains, snowflakes but not snow, branches but

one fist
five fingers

ten pairs
ten triples
five quads
one handful

not roots, heartbeats but not metabolism, airplanes but not clouds, notes but not music, words but not ideas. We cannot count things that change faster than our ability to assign numbers to them. We cannot count the discrete packets of data distributed over the internet because there are too many and they change too quickly. If we were to freeze the internet, or any physiological process, or any natural process, we might be able to count the static pieces that break off but we would be counting only the fragments of a dead thing, the living thing would have moved on to be something different. We cannot accurately say out loud the national debt (even if we could count the dollars) because by the time it was out of our mouths, the debt would have changed by a personal fortune.

There is no 3 in our physical reality, only 3-of.[8] Every object/interrelation within the physical world is unique and entangled. Uniqueness is a consequence of participation in a contextual web of interrelations. Counting then is a process of successively making and dissolving distinctions, with the goal of removing all uniqueness and discernibility from an interaction with physical reality.

In his development of **semiotics**, C. S. Peirce identified three types of abstract reference. An **icon** indicates by *looking like* what it represents. Delimiting brackets are an iconic representation of containers. An **index** tallies, it is *connected to* what it represents, it draws attention but does not resemble. While being constructed, tally marks are indices. A **sign** is *arbitrary*, with no connection to what it represents other than convention. Words are signs.

Conventional counting (1, 2, 3,...) does not count objects, it is rather an *indexing* scheme. Richard Heck: "Counting... has nothing essential to do with equinumerosity (let alone one-to-one correspondence)."[9] Conventionally we assign an index to each object by adding 1 to the index of the prior object. We do not add 1 to the object itself, we add it to the running tally. Any difference will do for indexing,

Accumulation

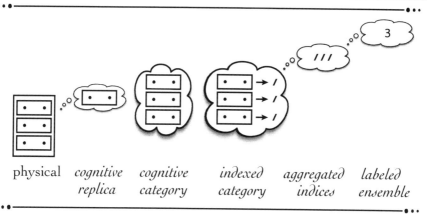

Figure 9-2: *Counting drawers*

---

numbers are only a convenience. The running sequence too has nothing to do with a tally, it is yet another cognitive scheme to keep track of thoughts. We do not assign n + 1 to a particular object, any one that remains unindexed will do.

The distinction between object and indicator is the **Fundamental Principle of General Semantics**.[10]

General Semantics
*Do not confuse the thing itself
with the tally mark that we have associated with it.*

The indistinguishability of tallies is essential to our ability to count them, otherwise the differences between them would overshadow the similarity that underlies their ensemble.

For the collection of furniture, we can assign the number 1 to the table, or to any of the chairs, or to any of the items. Similarly for 2, 3, etc. Clearly there is no preferential *"Item 3"*. The item that receives the number 3 can change if we were to count again, even though the total number of items does not change. Figure 9-2 provides a visual example. If you have ever watched a toddler learning to count, you will see the profound confusion caused by our arbitrary assignment of sequential numbers to objects.[11]

Chapter 9

	COGNITION
identification	Identify some unique objects (make a cognitive distinction to sever reality)
categorization	Construct a communality across objects (fuse some cognitive distinctions)

	COMPUTATION
indication	Associate an abstract indicator with each object (index members of a category)
fusion	Aggregate indicators (collect indices into the same container)
counting	Assign a cardinality to the fusion (label the container of the indices)

Figure 9-3: *The choreography of counting*

Choreography

only humans count

Figure 9-3 shows the choreography of counting. We first perceive and delineate replicas by making distinctions to separate our desired focus from the cacophony of reality.[12] Identification is an act of consciousness. The distinctions we impose to generate our object-replicas remove at the very beginning the idea that we are counting something in physical reality. We then ignore sufficient details and interrelations of replicas to construct a cognitive category. These steps are psychological rather than mathematical.[13]

only symbols can be counted

In order to count, we then enter a computational phase. We indicate each replica within our constructed category and then collect those indications. The categorization is maintained by forming a frame. Finally we assign a label (a number) to the tally of indicators within the frame.

This choreography does not call upon sequential counting because sequential counting is a *methodology* of assigning cardinality (at the last step), not a fundamental property of accumulation. Abstract indicators can be replicated precisely because they are virtual, not of the physical world. During the transition from cognition to computation we strip away sufficient detail from objects to consider them in some way equivalent to their tally.[14] To arrive at an integer count N, we collect the impoverished tally marks within a common frame. We call the *cardinality* N of our framed tally marks the **number of objects**.

What is accumulated in counting is not objects themselves, nor is it objects grouped together by a common feature. We can accumulate and count only the associated indications of each object that we define by our selective perception. That is *why* arithmetic is abstract. Mathematical lore has natural numbers living in some kind of virtual reality, the Platonic realm, and indeed that is where they belong. To count, we reduce reality to replica, replica to category, category to framed ensemble, and ensemble to cardinal number.

Addition too is subject to context and application. Helmholtz, a pioneer in mathematical physiology, sought to correlate the addition of numbers with physical measurements (1887).[15] He concluded that the axioms of natural numbers do not characterize reality. We should instead choose metrics to fit natural circumstance. One cloud plus one cloud does not make two clouds, but rather one possibly larger cloud. One cup of alcohol plus one cup of water does not make two cups of liquid, but rather about 1.9 cups. Two molecules of hydrogen plus one molecule of oxygen make two molecules of water.[16] Davis and Hersh draw the obvious but mathematically controversial conclusion:

Hermann vonHelmholtz 1821–1894

> There is and there can be no comprehensive systemization of all situations in which it is appropriate to add. Conversely, any systematic application of addition to a wide class of problems is done by fiat.[17]

Thus it behooves us to look closely at how counting works.

Chapter 9

Domain

Conventionally the **domain of discourse** specifies the set of objects are we addressing. Within mathematical logic, the domain is specified outside of the axioms and transformation rules, as a limitation on the type of objects that the transformations apply to. For example,

every number has a successor

$$\forall n \quad n \in \mathbb{N} \Rightarrow n+1 \in \mathbb{N}$$

The universal quantification $\forall n$ asserts we are referring to all of the domain. The double struck $\mathbb{N}$ indicates that the rule applies to the domain of natural numbers only. $\mathbb{N}$ defines what we mean by "every number". The set membership statement $n \in \mathbb{N}$, declares that whatever the variable n indexes belongs to the set of natural numbers. The rule says that for any natural number, there is a natural number that is one unit larger. The final set membership statement, $n + 1 \in \mathbb{N}$, declares that the successor number $n + 1$ is also a member of the set of natural numbers. Not only does the rule define a valid transformation (we can add one to any number), it also generates all admissible numbers. Combined with other rules, it contributes to the definition of our concept of **natural numbers**.

open

closed

Closure is a fundamental mathematical idea asserting that when an operation is applied to members of a specific domain, then the result is also a member of that domain. Multiplication of natural numbers is closed, for example, because all such multiplications generate another natural number. Closure of James forms is trivial because the three axioms permit only transformation between valid finite container forms. There is no mystery about what the result of an operation may look like.

But there is a subtlety for void-based forms: closure loses meaning when transformations create void-equivalent forms. The *domain of non-existent forms* is a nonsense concept. Since void-equivalent forms are meaningless, their illusionary presence is irrelevant. We have the choice of building meaning upon a domain that includes a lot of meaningless junk, or building the domain upon meaning, in which case there will be meaningless junk to clear away. Stretching to adopt Gödel's perspective, if we wish to address all possible structures (a complete system), then some will be illusionary (an inconsistent system). Alternatively if we wish to avoid illusion (a consistent system), then some structures will transform into nothing, leaving the system incomplete.

A conventional mathematical approach to describe how boundary forms work would be to first construct a domain of all possible patterns of containment, the **language** of boundary forms. The James rules convert members of the pattern language into other valid members of the pattern language. Closure restricts the transformations themselves from generating or accepting forms that are not in the language. For example ([A)] is not a valid containment pattern.[18] Fortunately the pattern language of containment has physical realizations that clearly identify valid containment patterns. The above invalid pattern would require us to place a physical A in two different containers at the same time. Or it might require us to be able to construct physical containers that intersect one other, an impossible construction if we wish to maintain the integrity of physical boundaries.

not a distinction

The form ([A)] is possible as an *imaginary* pattern, but that extension to the concept of containment would also require a completely new mapping to the interpretation. That type of mapping is even difficult to write in conventional notation. Function notation also obeys the structural restrictions of containers. But we could read ([A)] as a Venn diagram, for instance, where the round- and square- boundaries each identify a set, and A identifies the intersection of both sets. The two overlapping functions become two different properties shared by the single object A. Alternatively, we might introduce the logical concept AND to be able to say that A is contained both by ( ) *and* by [ ]. From the perspective of containment, the concepts of overlap, intersection and logical conjunction are necessarily imaginary concepts, in the meta-language or possibly imported from a different mathematical system.

Indication

A **distinction** creates a cognitive object that we are calling a replica. A **category** is a cognitive collection of replicas. We index replicas by tallying. A **tally** is created by associating a replica with a single unit marker or *tally mark*, usually with the intended purpose of fusing the tally marks together. Tally marks can manifest as any discrete index, including typographical strokes, checks, crosses or dots, and including physical beans, beads, buttons or pebbles. The abstract indicator or tally can be physical but it stands in place of a replica, *not* in place of a physical object. Brian Rotman:

> We count by repeatedly enacting the elemental process of creating identity by nullifying difference, repeatedly affixing the same sign "1" to individual "things" — objects, entities — that are manifestly not the same qua individuals in the world-before-counting from which they have been taken.[19]

Without numeric interpretation, we can say that Indication builds a frame, ([A][o]), around each replica A to be counted. We can call this particular type of frame an **indication frame**. The form o is a tally mark indicating or signifying a replica.

Here is the Indication Theorem that lays the groundwork for the process of counting.

indication
unmark ⇌ *mark*

$$A = ([A][o])$$

The demonstration requires only Inversion.

A	
([A])	enfold
([A][o])	enfold

An indication frame assures that a tally is void-equivalent, and thus has no numeric value. This allows us to construct a tally within a boundary form without changing the forms being tallied. The template for an indication frame is

(void-indicator [replica])   **indication frame**

Here, the round- and square-brackets are distinctions, not numeric operators. The Indication theorem specifies that any form A can be indicated by a unit, o. The content-form A itself can be a pattern, a numeric abstraction or a replica. A is not limited to a particular domain. Indication gives permission to associate a mark with A, in effect changing its type from unique to unitary. Indication frames convert the essential diversity of A into the preferred identicality of o.

Replication

We'll need to augment physicality with multiple reference, the ability to indicate an object more than once. This permits us to construct containment patterns in which a representation can occur in multiple contexts. For example, ((A)[A]) is a valid pattern that references the form A more than once. In terms of Peirce's semiotics, we will need indices to refer to icons.

We have used the ellipsis informally to indicate a multitude of replicas of the object A. It will be convenient to be able to show the cardinality of such a collection in notation, so I'll introduce a convenience notation for N replicas, $..._N...$. The ellipsis notation is in the meta-language, it is not technically part of boundary algebra. However the labeled ellipsis does identify a property of the container. N indicates the cardinality of the indicators.

$$A..._N..A = ([A][o..._N..o])$$

replication
replicate ⇄ tally

Chapter 9

The Replication theorem connects a multitude of replicas to an equinumerous multitude of unit indicators, so that replicas and tally marks are constructed in one-to-one correspondence. The theorem is a consequence of Arrangement. Here is the demonstration.

	A ..ₙ.. A
mark	([A][o])..ₙ..([A][o])
collect	([A][o..ₙ..o])

Once tally marks have been collected within a single container, the act of counting objects becomes independent of the actual objects being counted. As an example, we'll tally some (icons of) flowers.²⁰

	❀ ..ₙ.. ✸
mark	([❀][o])..ₙ..([✸][o])
collect	([❀][o..ₙ..o])
label	([❀][N])

We begin with some flower icons. Each is different. We mark each icon with a tally mark, collect the tallies into a container, and then name that fusion of tallies by its cardinality N. To refer to the number of flowers abstractly, we keep the N and throw away the flowers. The individual personality of each flower is lost during the collect operation, not during labeling.

Note that *no sequential counting* has occurred. The assemble of tallies, of whatever cardinality, stands in place of the collection of flower replicas. We can name the cardinality N without knowing the value of N. We can say that the value of N is the collection o..ₙ..o.

You might be wondering how then can we tell the numbers apart? What makes 2 different than 3? Difference is failed correspondence. By placing tallies of different ensembles into correspondence, we can identify whether the ensembles have a different cardinality, still without introducing counting or sequence. By keeping track of which ensemble has remaining units after a failed correspondence, we *could* empirically put ensembles in order by magnitude, again without ever counting. That is, we can construct both ordinal sequences and cardinal magnitudes without counting.

At this point, we have now added two new types of frame, **indication** and **cardinality**.

$$A = ([A][o]) \qquad \text{indication frame}$$
$$A.._N..A = ([A][o.._N..o]) \qquad \text{cardinality frame}$$

We could also read the Indication theorem as a numeric operation, multiplying A by the unit 1.[21] When the indicator is greater than one, as in ([A][N]), the frame identifies a cardinality, the number of As. The cardinality frame can be interpreted, as well, as multiplication. More generally, the generation of cardinality does not need to look like a multiplication of a quantity by a quantity, for example ([❀][N]). Arrangement is not necessarily about multiplication, here it is about collecting tallies. Cardinality is a by-product, while multiplication is a limited interpretation. Thus we could generate an arrangement of flowers, ([❀][❀][❀]), for which flowers index other flowers without an implication of numerosity.

Generalized Counting

The three separate computational processes that contribute to generalized counting are marking, fusing, and naming. Accumulating tally units (i.e. counting in parallel by 1s) is the simplest case of indicating cardinality. Frames apply more broadly. Counting-by-ones relies upon the collection and fusion of *unit* tallies. We can *count-by-twos* (more generally by Ns) in at least three different ways:

— by fusing pairs of tallies,   ([❀][oo])
— by marking pairs of replicas with a single tally   ([❀❀][o])
— by reading each individual tally as oo (i.e."2").   o ☞ 2

Chapter 9

The common interpretation of counting "two-at-a-time" is the sequential accumulation of pairs of tallies. We would count 2, 4, 6,... rather than 1, 2, 3,.... This simple numeric description masks the necessary subprocesses of grouping (by twos), collecting groups with (dual) tally marks, and collecting again the paired tallies. These subprocesses are illustrated below. I'll use flower icons as objects.

```
                          ✿  ..N..  ✿
mark             ([✿][o])    ..N..   ([✿][o])
enfold    ([([✿][o])([✿][o])])..N/2..([([✿][o])([✿][o])])
collect        ([([✿][oo])])..N/2..([([✿][oo])])
clarify         ([✿][oo])   ..N/2..  ([✿][oo])
collect              ([✿][oo..N/2..oo])
```

Counting by N accumulates N replicas over M counts, resulting in N x M marks. Counting by N is multiplication by successive addition. This type of counting is built deeply into our numeric system as making groups of the same magnitude (i.e. a common base). We use base-10, so we make groups of 10 out of ten 1s, and then groups of 100 out of ten 10s, etc. 500 is counting by 100, five times. In the process of tallying the groups above, collect is the key operation. The Arrangement axiom itself embeds an interpretation of multiplication into James forms.

We can also count pairs of replicas (such as married couples, pant legs, eyes and ballroom dancers) directly by associating each pair with a single tally mark.

```
                     ✿    ..2N..   ✿
enfold        ([✿✿]    )..N..([✿✿]    )
enfold        ([✿✿][o])..N..([✿✿][o])
collect            ([✿✿][o..N..o])
```

An example of this type of counting would be counting pairs of shoes, but not individual shoes. We must still recognize the subprocess of grouping shoes by pairs in the first place, but we do not need to construct the sequence 2, 4, 6,....

Each pair ❀❀ is associated with an indication frame via the structure ([❀❀][o]). The unit tallies are then collected to yield a single form that identifies both the pair and the number of occurrences of the pair. Here is where multiplication intersects with counting. The number of times that ❀❀ is replicated becomes the number by which ❀❀ is multiplied. Counting objects two-at-a-time localizes pairs by manipulation of tallies. Counting pairs of objects one-at-a-time localizes pairs to the cognitive steps prior to counting, as construction of compound objects.

The final form of counting-by-twos is rarely recognized because of its perversity. We could elect to use the sequence 2, 4, 6,… rather than the sequence 1, 2, 3,… to indicate counting by ones. This of course is both a syntactic trick and confusing. We are in effect saying that the distance between each unit is 2 rather than 1. This strange behavior is worth bringing to attention because it points to a conceptual oversight. When we think of the symbols "1", "2", "3" simply as meaningless names for the items in a sequence, then the sequence of symbols "2", "4", "6" conveys the same intent. The *by-twos* sequence has the same numeric characteristics as the *by-ones* sequence but for a scaling factor. Each set of symbols forms a well-ordered sequence with equidistant indicators. Both start from the same zero. We have simply redefined the distance between digits.

Different sequences provide the same services as good old 1, 2, 3,… which means that sequential counting is conceptually ambiguous. To be unique, Peano's successor axiom, $n \Rightarrow n + 1$, must also identify the distance between unit steps, that is, the size of the unit. Indeed, any linear function satisfies the successor function upon which Peano built his axioms of arithmetic. The definition of numbers and the rules of arithmetic work just as well with any scale unit or with any uniformly increasing scale. We have built our numbers on shady foundations, and a direct consequence is that sequential counting is more complicated than counting tallies 1, 1, 1,….

9.3 Multiplication

The Arrangement axiom defines the general *process of collection* (right to left) without regard to cardinality and without regard to what is being collected. It describes the general *process of replication* (left to right), again without regard to the number of replicas created or to what is being replicated.

arrangement   (A [B...Z]) = (A [B])...(A [Z])

Tallying and replicating are special cases of Arrangement in which the forms B through Z are tally marks and the replicated forms are indication frames ([A][o]). As an example, 2 x 3 matches the pattern of Arrangement with A = [oo].

```
            ([oo][ooo])
disperse    ([oo][o]) ([oo][o]) ([oo][o])
unmark         oo         oo         oo
```

With A = [ooo] arrangement proceeds differently, although the transformation steps are the same.

```
            ([ooo][oo])
disperse    ([ooo][o]) ([ooo][o])
unmark         ooo         ooo
```

The iconic form ([A][N]) can be interpreted as "collect A, N times". Conventionally, we think of A as a group, say a group of ten, and N as the number of groups. This constructs an asymmetry. The group remains an ensemble, while another form, the N here, is partitioned into units each of which indexes a newly constructed replica of the ensemble. But multiplication is more complex still since both group/ensemble and replicator/indexes need not be whole numbers.

Arrangement does not specify an implementation strategy, although the form of the axiom suggests parallel processing. [A] does not "precede" [N], regardless of appearance. Thus we could read ([A][N]) as "N copies of A" or as "A copies of N". More generally, the iconic form of multiplication represents any collection of repeated processes that replicate any context. This conceptualization of multiplication makes a clear distinction between what is doing the multiplying and what is being multiplied. 3 x 5 is interpreted conventionally in two different ways. Does it mean 5 + 5 + 5 or does it mean 3 + 3 + 3 + 3? Conventional multiplication uses argument ordering to identify multiplication *of* one item *by* another item.

addition is about replicas

multiplication is about relations

James multiplication has no ordering principle. However what is being dispersed and what is being replicated are clearly distinguishable. The James form differentiates one container as being partitioned and dispersed (multiplying by) and any others in the same context as being replicated (multiplication of), while the implementation specifies which is which.

multiplication
([of][by])

Multiplication is usually considered to be shorthand for the accumulated additions.

$$A \times N = A + .._N.. + A \; \mathbb{F} \; A.._N..A$$

multiplication

The Replication theorem converts between multiplication and addition by decomposing a cardinality N into units, and then dispersing those units to construct N replicas of the reference to object A. Finally the references are laid out in an array.[22]

$$A \times N \; \mathbb{F} \; ([A][\quad N \quad])$$
$$([A][o.._N..o])$$
$$([A][o]).._N..([A][o])$$
$$A \; ..\_N.. \; A \quad \mathbb{F} \quad A + ..\_N.. + A$$

substitute
disperse
unmark

Multiplication is most certainly *not repeated addition*. Addition operates upon the same type of object. Counting also acts upon the same type of object by degrading difference into tally marks. Multiplication acts upon *different types* of objects. Multiplication pays attention to packaging, as is illustrated by its form, ([A][B]).

Multiplication becomes repeated addition when replication replaces structure. For example, ([✾][✾]) is Cartesian multiplication whereas ([✾][N]) is multiplication-as-addition. Multiplications accumulate, just like additions. The form of repeated multiplication resembles the form of repeated addition, but with the characteristic configuration of square-bracketed forms contained by a single round-bracket.

A x B x ... x Z ☞ ([A][B]...[Z])

ranks x *suits*

We can multiply replicas without numerics by forming a **Cartesian product**. A deck of playing cards, for example, is composed of the product of four suits by thirteen types. If we restrict the example to royal cards (the King, Queen, and Jack), then the product is ranks by suits, with no mention of numbers. At the level of unit types we can factor the *ranks* x *suits* array into the form of non-numeric multiplication:

([♠ ♥ ♦ ♣][K Q J])

Here's an example of generic (Cartesian) multiplication.

([♠ ♥ ♦ ♣][K Q J])

disperse   ([♠ ♥ ♦ ♣][K]) ([♠ ♥ ♦ ♣][Q]) ([♠ ♥ ♦ ♣][J])

disperse   ([♠][K]) ([♥][K]) ([♦][K]) ([♣][K])
              ([♠][Q]) ([♥][Q]) ([♦][Q]) ([♣][Q])
              ([♠][J]) ([♥][J]) ([♦][J]) ([♣][J])

The results of dispersion remain in the form of "multiplication", indicating that each component form has two specific properties, one rank and one suit.

Multiple applications of Arrangement expose the relation between the Additive and the Multiplicative Principles.

distribution of addition over multiplication

([A B][C D]) = ([A][C]) ([A][D]) ([B][C]) ([B][D])

This relationship is conventionally described as *distribution of addition over multiplication* or as conversion between factored and polynomial expressions. Arrangement is broader.

9.4 Exponents

What if we want to *multiply* 2 by itself three times? How is multiplication itself replicated? Just like the fusion of tallies (i.e. identifying the cardinality of an ensemble of units)

$$1 + 1 + 1 \, \text{☞} \, o \quad\quad o \quad\quad o$$
$$([o][o])([o][o])([o][o]) \quad\quad \text{mark}$$
$$([o][ooo]) \quad\quad \text{☞} \, 1 \times 3 \quad \text{collect}$$

and just like the fusion of groups (i.e. factoring the cardinality of an ensemble)

$$2 + 2 + 2 \, \text{☞} \, oo \quad\quad oo \quad\quad oo$$
$$([oo][o])([oo][o])([oo][o]) \quad\quad \text{mark}$$
$$([oo][ooo]) \quad\quad \text{☞} \, 2 \times 3 \quad \text{collect}$$

We can aggregate repeated multiplications by marking and collecting them.

$$2 \times 2 \times 2 \, \text{☞} \, ([oo][oo][oo])$$
$$(([[oo]][o])([[oo]][o])([[oo]][o])) \quad\quad \text{mark}$$
$$(([[oo]][ooo])) \quad\quad \text{collect}$$

Repeated addition generates the form of multiplication

$$B + B + B = B \times 3 \, \text{☞} \, ([B][ooo])$$

More abstractly,

$$B + .._N.. + B = B \times N \, \text{☞} \, ([B][o.._N..o]) = ([B][N])$$

Repeated multiplication generates the exponential form.

$$B \times B \times B = B^3 \, \text{☞} \, ([B][B][B]) = (([[B]][ooo]))$$

The abstract form of exponentiation as repeated multiplication is

$$B \times .._N.. \times B = B^N \, \text{☞} \, ([B].._N..[B]) = (([[B]][N]))$$

Repeated units take the form $o.._N..o$, repeated additions take the form $A.._N..A$, while repeated multiplications take the form [B]..ₙ..[B]. Units are an ensemble of ground forms (i.e. empty containers), sums are an ensemble of variable forms, and multiplications are an ensemble of logarithmic forms with the label B standing in place of the base.[23] In James algebra then, we have the same specific interpretation for the operations of counting, multiplying, and raising to a power. These apparently different operations are a single operation nested successively within itself.[24]

We have interpreted the behavior of the round boundary as converting its contents into an exponent with an indeterminate base.

$$(A) \quad \mathrel{\text{☞}} \quad \#^A$$

The James form of exponentiation, (([[B]][A])), constrains the base-free notation to a specific base-B.

$$((\lbrack\lbrack B\rbrack\rbrack\lbrack A\rbrack)) \quad \mathrel{\text{☞}} \quad B^A$$

We can carry out successive multiplications of the arbitrary base itself.

$$\#^N = \#\text{x}..\text{}_N..\text{x}\# \quad \mathrel{\text{☞}} \quad ([\#]..\text{}_N..[\#]) = ((\lbrack\lbrack\#\rbrack\rbrack\lbrack A\rbrack))$$

is an abbreviation for the doubly-nested round-bracket,

$$\# \quad \mathrel{\text{☞}} \quad (\text{o}) = ((\ )) \quad \mathrel{\text{☞}} \quad \#^1 = \#^{\#^0}$$

The two forms of exponentiation are consistent:

	$\#^A \quad \mathrel{\text{☞}} \quad (([\lbrack \quad \# \quad]][A]))$
substitute	$(([[((\ ))]][A]))$
clarify	$(\qquad\qquad A \ )$

Rules

The result of raising any base to the zero power, $\#^0$, and the result of raising any base to a negative power, $\#^{-N}$, are usually given as definitions rather than as the consequences of structure. The James axioms show that these definitions are deducible theorems. Below, the hash-mark has been replaced by the variable letter B to indicate that we are limiting the base of the power form to be a natural number other than 1. Since the concept of a negative number has not yet been introduced, we'll postpone discussion of negative exponents (i.e. unit fractions) until Chapter 10.

It is easy to show that $B^0 = 1$, and that $B^1 = B$.

```
B⁰ ☞ (([[B]][ ]))
     (           )       ☞ 1              absorb


B¹ ☞ (([[B]][o]))
     ( [B]      )                         unmark
            B           ☞ B               clarify
```

James forms replicate the behavior of conventional exponents without the introduction of a special class of transformations, the conventional rules of exponents. The next three demonstrations show that the rules of exponents (rearranging, adding and multiplying) are structural variations that require only Arrangement and Inversion.

```
Aᴺ x Bᴺ ☞
     ([ ((([[A]][N])) ][ ((([[B]][N])) ])
     (   ([[A]][N])      ([[B]][N])    )    clarify
     (   ([ [A][B] ][N])                )   collect
     (   ([[([A][B])]][N])              )   enfold
                        ☞ (A x B)ᴺ


Bᴹ x Bᴺ ☞
     ([ ((([[B]][M])) ][ ((([[B]][N])) ])
     (   ([[B]][M])      ([[B]][N])    )    clarify
     (   ([[B]][M N])                  )    collect
                        ☞ Bᴹ⁺ᴺ


(Bᴹ)ᴺ ☞
     (([ ((([[B]]  [M])) ]][N]   ))
     ((     [[B]]  [M]       [N] ))         
     ((     [[B]][([M]     [N])] ))         clarify
                                            enfold
                        ☞ Bᴹˣᴺ
```

Chapter 9

operation	iteration	form	replica	replication	interpretation
count	o..$_N$..o	([o][N])	o	N	1 x N
add	o..$_M$..o o..$_N$..o	([o][M N])	o	M+N	1x(M+N)
multiply	B..$_N$..B	([B][N])	B	N	B x N
power	([B]..$_N$..[B])	(([[B]][N]))	[B]	N	B^N

Figure 9-4: *Accumulating and grouping tallies*

Notice also that in both of the above applications of enfold, the boundary form is made more complicated in order to match the structure of the conventional interpretation.

Logarithms

As is well known, multiplication can be achieved by addition of logarithms, followed by back-conversion via an **antilogarithm**, or power, function. This is how a slide-rule, for example, achieves multiplication. To use the logarithm finesse in multiplying 2 by 3, we could add log 2 and log 3 together, then take the antilogarithm of the sum.

$$2 \times 3 = \#^{log\#2} \times \#^{log\#3} = \#^{log\#2 + log\#3} \quad \text{☞} \quad ([2][3])$$

The final James form is literal and can be derived directly by expressing conventional exponential notation as a string of characters. The form of 2 x 3 is recorded below as a string, with the arguments of each operation enclosed in a boundary.

$$2 \times 3 = \#^\wedge(log_\#[2] + log_\#[3])$$

absorb power, log   2 x 3 = ([2] + [3])
absorb addition   2 x 3 = ([2] [3])

The James form first absorbs the symbolic exponent and logarithm operations into the two different brackets. Then addition is absorbed as shared content of the round-bracket.

9.5 The Accumulation Family

Figure 9-4 provides a summary of how we accumulate tallies. There is essentially only one mechanism, collecting identical forms into the same container. Figure 9-5 is an extended example of the process of counting. It begins with the objects being replicated, in this case flowers: ❀, ❀, ❀. The cognitive steps of breaking each object out of its web of associations to form replicas, and of categorizing replicas as flowers to form flower-frames, have already been done before the start of arithmetization.

Indication is the first step toward counting. Each replica in the flower category is marked with an indicator, or index. The indicators (tallies) are then collected and the collection named with a number symbol, the final step in what we call *counting*. The Additive Principle depends upon this loss of unique identity.[25] We can add only tallies. In Chapter 2 this process was described as **fusion of ensembles**.

Once we move to multiplication, accumulation loses its grounding in tally marks. The difference between addition and multiplication is that indicators are added, while categories are multiplied. For addition, the flower frame, ([❀][*framed-form*]), is factored out via Indication. Flowers no longer participate in the addition. For multiplication, the flower frame remains as an active participant in the multiplication.

([❀][❀]) is certainly a valid containment form, however, the numeric interpretation loses meaning. The general principle is that containers are domain-independent, but interpretations are not. With the mixed iconic/symbolic form, ([❀][5]), we can accumulate five flowers but not flower fives. We experience a forced asymmetry of cognitive and numeric. The Additive Principle permits decomposition of the numeric 5 but not the category ❀.

Chapter 9

To Indicate an Object
construct a framed unit (and lose physical reality)

1	☞ ([✻][o])	*mark*
	1	count

To Indicate Cardinality
collect units into the same frame (and lose uniqueness)

3	☞ ([✻][o]) ([✻][o]) ([✻][o])	*mark*
	([✻][ooo])	collect
	3	count

To Add
collect ensembles that have the same frame-type (and lose origins)

3 + 2	☞ ([✻][ooo]) ([✻][oo])	*tally*
	([✻][ooooo])	collect
	5	count

To Multiply
disperse ensembles inside the same frame (and lose objects)

3 x 2	☞ ([✻] [ooo][oo])	*cross-tally*
	([✻][([ooo][oo])])	enfold
	([✻][([ooo][o])([ooo][o])])	disperse
	([✻][ooo ooo])	unmark
	6	count

To Raise to a Power
disperse collected multiplications (and change contexts)

2³	☞ ([✻]([[oo]][ooo]))	*power-tally*
	([✻]([[oo]][o])([[oo]][o])([[oo]][o]))	disperse
	([✻] [oo] [oo] [oo])	unmark
	([✻][([oo] [oo] [oo])])	enfold
	([✻][([oo][oo][o]) ([oo][oo][o])])	disperse
	([✻][([oo][oo]) ([oo][oo])])	clarify
	([✻][([oooo][oo])])	collect
	([✻][([oooo][o]) ([oooo][o])])	disperse
	([✻][oooo oooo])	unmark
	8	count

Figure 9-5: *James algebra calculation*

COGNITIVE CONSTRUCTION

REALITY ☞ replica

replica$_a$ replica$_b$ replica$_c$ ⇒ category

ARITHMETIZATION

$\{a, b, c\}_{\text{category}}$

$\{a_o, b_o, c_o\}_{\text{indexed category}}$

$(ooo)_{\text{ensemble}}$

$N_{\text{cardinality}}$

Figure 9-6: *Numeric abstraction*

Exponentiation, the arbitrarily maximal process in the accumulation hierarchy, appears to accumulate multiplications However, we will see in Volume III that the exponential context introduces a remarkable diversity of new mathematical creatures.

9.6 Remarks

We have been describing the arithmetization of reality. Mathematics, even natural numbers, is not part of physical reality, it is instead a cognitive artifact. Figure 9-6 reiterates the schematic process of *removing detail from reality in order to generate numbers*.

The first step in arithmetization is indeed construction of a set, but a set of cognitive replicas, distinctions that constitute instances of a concept. We then immediately depart from the unique elements of a set by constructing indicators that stand in place of each element. These indicators can be accumulated to form an ensemble, and only then can we take the step of assigning a cardinality. These steps trace Peirce's categorization of the

Chapter 9

elements of representation. Replicas are icons, icons are indexed, indices are converted into symbols that identify cardinality.

The two James transformation patterns are sufficient for defining the structures and operations of arithmetic. We have explored three primary numeric frames that will also suffice for the remaining concepts of arithmetic yet to come.

indication/addition	([B] [o])
cardinality/multiplication	([B] [N])
depth/power	(([[B]][N]))

A primary change of perspective is to see both numeric form and computation as parallel processes, all conceivably occurring at the same time. We thus can remove the last vestige of time and sequence from the numeric Inductive Principle. In Volume II we further explore the concept of concurrency, taking steps all at the same time by abandoning induction.

In general, the concepts of arithmetic also apply to non-numeric forms. Addition provides collection via fusion; tallying provides numeric counting; multiplication provides structural relation. Modern math education addresses magnitude at the complete exclusion of category. This might be an artifact of the twentieth century attempt to remove human cognition from the concepts of mathematics, similar to viewing trees as lumber rather than organisms. With categorical multiplication, we can begin to battle one of the costly mistakes of elementary math education: the belief by an overwhelming majority that math is about numbers.

Endnotes

1. **opening quote:** E. Mach (1895) *Popular Science Lectures*, T. McCormack (trans.) p.195-96.

2. **counting works through — it is — significant repetition:** B. Rotman (1993) *Ad Infinitum* p.6.

3. **quite independently of any connections with counting:** R. Heck (2000) Cardinality, Counting, and Equinumerosity. *Notre Dame Journal of Formal Logic* 41(3) p.200.

4. **both built numbers from fusions rather than sets:** M. Potter (2004) *Set Theory and its Philosophy* p.23.

5. **who have not internalized this set of ideas and made it intuitive:** P. Davis & R. Hersh (1981) *The Mathematical Experience* p.395.

6. **and decidedly not part of the natural world:** Kronecker made his famous quote *"God made the integers, all else is the work of man."* at a meeting in 1886 (from his obituary, 1823-1891, in *Mathematische Annalen* 43:1-25). It was recently publicized as the title of a book by Stephen Hawking, *God Created the Integers: The mathematical breakthroughs that changed history* (2007). The quote is associated with Kronecker's belief that mathematics should address finite numbers and should avoid the irrational numbers, a position similar to that taken here with regard to real numbers. But there is no need to interject metaphysics, man created the rationals as well as the irrationals. Brian Smith (*On the Origin of Objects*, 1996, p.355) comments that Kronecker got it backwards. Metaphysics arises out of indefiniteness, the absence of boundaries, from which we then carve out the discrete numbers. Irving Stein (*The Concept of Object as the Foundation of Physics*, 1996, abstract) points to the first distinction: "Objects arise from restrictions on nonspace."

Fields medalist Michael Atiyah draws the analogy of a jellyfish living in an ocean that provides perceptual continuity but not discrete objects. From an online interview 12/2016 at http://www.webofstories.com/play/michael.atiyah/89

> We see individual objects and if the universe were different and we didn't have individual objects around us we just had a sea, a mass of continuity…counting would not be natural.

Chapter 9

The **systems viewpoint** is that the web of interrelations between things is closer to reality than the things we create via perceptual discrimination.

7. detach the object from its web of interrelations and see it as a discrete thing: These distinctions are important because the epistemology that generates counting is no longer adequate for understanding and acting responsibly in the world. Our world is a *complex network of interrelations*. Physical objects do not have an independent existence, we carve them out of the web of reality by making cognitive distinctions.

We will not be able to reach a necessary holistic perspective on our Earth, we will not have access to ecological thinking with our current object-oriented perception of how to count what counts. Quantum mechanics leads the way to understanding the essential indeterminism of objects. Non-linear dynamics reinforces the idea that reality is a network of interdependencies. We look at a forest and see individual trees, unable to comprehend that 2/3 of each tree is below ground. It is the dense interweaving of roots that is the forest's cognitive network. In essence, counting destroys the richness of reality.

8. no 3 in our physical reality, only 3-of: For example, every rabbit is unique. When we see a rabbit on the lawn, our perceptual system constructs an experiential replica. Facets of that replica fit categories we have established from prior experience. We have in mind one replica of our personal category *rabbit*. Now we might look closer and see three rabbits. The cardinality of that ensemble can be assigned a number, the number of rabbits. However, if we introduce a different (imposed) perceptual category, say age, we cease to have three rabbits and may have instead one parent and two children. Cardinality depends upon not only a cognitive category but also the ability to hold that category consistent in a dynamically changing perceptual environment.

Similarly, *addition does not apply to abstract numbers*. Only 2-of and 3-of can be added together. This viewpoint is virtually opposite to the idea that mathematical operations are independent of application. The sum of the concept 2 and the concept 3 is a symbolic definition, not an operation in the sense of *put together*. Technically, the set of two $\{o_1, o_2\}$ put together with the set of three $\{o_3, o_4, o_5\}$ is a new set with two members, $\{\{o_1, o_2\}, \{o_3, o_4, o_5\}\}$. In order to get to five, the two sets need to be joined by Union; addition over sets, not addition over elements. To apply the Union operation, members need to be indexed so that they can be uniquely identified, since Union does not support duplicates. Addition as fusion avoids these issues.

9. **equinumerosity (let alone one-to-one correspondence):** R. Heck, p.202.

10. **the fundamental idea of general semantics:** Alfred Korzybski (1933) *Science and Sanity: An introduction to non-Aristotelian systems and general semantics.*

11. **caused by our arbitrary assignment of sequential numbers to objects:** This confusion may be at the root of our societal choice to educate young children to abandon their sense of number in favor of mimicking arbitrary activities dictated by elementary school teachers, the great majority of whom suffer from math anxiety.

12. **making distinctions to separate our desired focus from the cacophony of reality:** I'm avoiding the opportunity for a deeper philosophical exploration of the nature of perception here by naming our cognitive contents replicas rather than "objects".

13. **These steps are psychological rather than mathematical:** Mathematics classically eschews its cognitive origins, placing numbers, for example, within a separate Platonic reality. Delusion and metaphysics do not excuse accountability.

14. **to consider them in some way equivalent to their tally**: To put elementary arithmetic, counting, at the foundation of our educational system is an extreme of human arrogance. Imagine! First we sever the Whole, and then we deny the uniqueness of the severed parts so that we can tally the accumulation of our own destruction.

15. **correlate the addition of numbers with physical measurements (1887):** H. vonHelmholtz (1887) *Counting and Measuring.*

16. **Two molecules of hydrogen plus one molecule of oxygen make two molecules of water:** $2\,H_2 + O_2 \rightleftarrows 2\,H_2O$.

17. **application of addition to a wide class of problems is done by fiat:** Davis & Hersh, p.74.

18. **([A]) is not a valid containment pattern:** We are considering here only the act of distinction. A boundary identifies a difference. Set intersection denies that difference.

Chapter 9

19. **world-before-counting from which they have been taken:** Rotman, p.51.

20. **we'll tally some (icons of) flowers:** The idea that we are tallying *icons* is important. Icons and images are replicas, not originals. They can be counted directly since they are their own indicators. Icons stand between images and tallies, they look like flowers but are not images of actual flowers.

21. **a numeric operation, multiplying A by the unit 1:** The ancient Greeks did not consider 1 to be a number, rather One is unity, a whole. "Multiplying" by One is an absurd action, since multiplication is replication, not unity. This is another example of how the symbolic disembodiment of arithmetic has removed us from common sense.

22. **the references are laid out in a conventional array:** A subtle change has been introduced. A no longer labels a collection of objects that is being tallied, rather it labels another ensemble. *Numeric* multiplication is necessarily divorced from a physical referent; it is a relation between two cardinalities. In that sense, multiplication is *not* repeated addition. The mathematical conceptualization of multiplication is that of a **product space**, a single collection of all the possible ways to pair members of two separate collections. We have metaphorically characterizing the Multiplicative Principle by saying that every form in one container *touches* every form in another container.

23. **with the label B standing in place of the base:** We could iterate exponentials to form stacks of powers, and this is exactly how we would interpret deeply nested round-brackets.

$$((((A)))) \quad \Rightarrow \quad \#\#^{\#^{\#^A}}$$

24. **a single operation nested successively within itself:** This is the way that adding, multiplying and exponentiation are commonly taught to school children.

25. **The Additive Principle depends upon loss of unique identity:** An *additive system* exists only when its elements are interchangeable and indistinguishable. When permutations and combinations of elements matter, as is the case for organic systems, then a system is multiplicative. For application to economy, biology, psychology, cognition, and evolution, see G. Bateson (1972) *Steps to an Ecology of Mind*, p.358 and particularly the essay The Role of Somatic Change in Evolution, p.346.

Chapter 10

Reflection

The arithmetic operation of negation corresponds to the geometric idea of reflection.[1]
— *Paul Lockhart (2012)*

We have introduced two types of empty boundary, that is, two types of unit, two grounds. ( ) behaves like One, the conventional *unit*. [ ] behaves like negative infinity, an unconventional unit that we have thus far designated as undefined during transformation.

We now add a third and final type of boundary, the angle-bracket < > to represent the concept of *reflection*. The James angle-bracket can be interpreted as a *generalized inverse*. Since we have built cardinality, counting, addition, multiplication, and exponentiation out of the same two boundary containers, any new container will apply equally to all of these interpretations. That is, we need only one more *reflection boundary* to define counting backwards, negative numbers, subtraction, division, logarithms and taking a root. The minus sign that identifies negative numbers is but one instance of a more general concept. Conventional notation also makes the leap of associating negation with division in the expression $A^{-1} = 1/A$.

< > *provides inverse operations*

Chapter 10

The James form of the **negative unit** −1 is a composite of the positive unit, o, and the reflection operation, <o>. When applied to ensembles, angle-brackets generate the negative numbers. When A is a positive number, <A> is its negative inverse.[2]

numeral	James form
−0	< >
−1	<o>
−2	<o o>
−3	<o o o>
−N	<o..$_N$..o>

reflect

The word *reflection* is intended to elicit a physical metaphor. Just like a mirror, the reflection of a concrete object generates an imaginary image. In fact, *all reflections are imaginary*. Geometric reflection is the fundamental operation of movement in a two-dimensional plane. Both translation and rotation can be expressed as a composition of two reflections. Analogously, the inverse operations of arithmetic can be expressed by the fundamental operation of boundary reflection.[3]

rotate

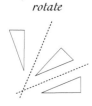

Since a reflected image is dependent upon an original form, the angle-bracket is almost always read as an *operator*. Both the form that is reflected and the context of reflection determine the interpretation of the reflection operation.

translate

As a type of boundary unit, (i.e. as a container of nothing), < > cannot be interpreted as a number. < > cannot reflect *void* since *void* does not exist. We can therefore take the empty angle-bracket to be void-equivalent.

$$< \ > \ = \ void$$

Void Reflection has a useful corollary: all void-equivalent forms are blind to reflection.

void reflection   A = <A>   *when*   A = *void*

10.1 Patterns

Forms and reflected forms in the same container can be *mutually deleted*. We'll take the structure of **Reflection** as an axiom that defines the angle-bracket.

$$A <A> = \text{void}$$

reflection
create ⇄ cancel

Reflection is a void-equivalent pattern. It provides another technique for identifying structure that can freely be discarded. The void-equivalence of the unit < > is now simply a special case of Reflection, for which A is *void*. Here's Reflection within an explicit outer container.

$$(A <A>) = (\ )$$

The Reflection axiom leads immediately to several simple theorems for moving angle-brackets around, and two essential theorems, Dominion and Promotion, that connect the three bracket types together. But first the simple theorems.

The **Involution** theorem asserts that the angle-bracket is its own reflection.

$$<<A>> = A$$

involution
wrap ⇄ unwrap

We can demonstrate Involution quite easily,

```
<<A>>
<<A>> <A> A          create
         A           cancel
```

Separation is another simple theorem for angle-brackets that arises directly from Reflection.[4]

$$<A><B> = <A\ B>$$

separation
split ⇄ join

Here's the demonstration.

```
<A><B>
<A><B><A B> A B      create
       <A B>         cancel
```

The **Reaction** theorem exchanges the reflection boundary across forms.[5]

reaction

$$\langle A \langle B \rangle\rangle = \langle A \rangle\ B$$

Again, a simply demonstration.

	< A <B> >
wrap	< <<A>><B> >
join	< <<A> B> >
unwrap	<A> B

Dominion

Dominion is the most important theorem in James algebra since it is a powerful tool for identifying void-equivalent forms. The Dominion theorem does not include an angle-bracket, but depends upon Reflection for its proof. This is our first hint that angle-brackets are needed now but will be shown to be superfluous later.[6]

The Dominion Theorem generalizes the Unification axiom for square-brackets. Any numeric form that is in the same container as [ ] is absorbed. Dominion asserts that (A [ ]), a *frame without a framed-content*, is void-equivalent. Empty frames are illusions.

dominion
emit ⇌ absorb

$$(A\ [\ ]) = \textit{void}$$

This void-equivalent theorem is the ground case of the mechanism that permits replication of structure without changing cardinality. It permits the introduction of any numeric form A in the context of the non-numeric form [ ], thus providing our first perspective on the nature of [ ]. The square-bracket emits indeterminate numeric forms.

Dominion is the base case of cardinality.

([A][ ])	*cardinality* 0
([A][o])	*cardinality* 1
([A][N])	*cardinality* N

Reflection

Dominion clearly mixes numeric with non-numeric forms, making it a chimera from the perspective of conventional expressions. One interpretation of Dominion is that of multiplication by zero.

$$A \times 0 = 0 \quad \text{☞} \quad ([A][\ ]) = \textit{void}$$

Under this interpretation, it is clear that A can be any numeric form.[7] Dominion can also be interpreted as raising a numeric base to the power of 0.

$$B^0 = 1 \quad \text{☞} \quad (([[B]][\ ])) = (\ )$$

[[B]], the base of the exponential expression, is absorbed. Since the form of B is deleted by Dominion, we can immediately generalize that any numeric expression raised to the zero power is 1. More fundamentally, within the James notation B is void-equivalent and thus *non-numeric*.

$B = ((\ ))$
$[[B]] = \textit{void}$

Yet another interpretation of Dominion is the sum of a numeric expression A with negative infinity [ ], coupled with the idea that dividing by infinity results in 0.

$$(A\ [\ ]) \quad \text{☞} \quad \#^A \times 0 = 0$$
$$\text{☞} \quad \#^{A+-\infty} = \#^A \times \#^{-\infty} = \#^A / \#^\infty = 0$$

Usually this observation is substantiated by calling upon the theory of limits. But [ ] is a foundational James unit, not a special case. Numeric forms (numbers) and non-numeric forms (infinities) are incommensurable, but they do, to a limited degree, interact. Dominion is the pattern that bridges the two worlds.

In Dominion we have a single explanation for a diversity of non-numeric expressions, and a start to knitting numeric and non-numeric forms into a coherent theory of extended arithmetic.

$$\begin{array}{lll}
A\ [\quad\quad\ ] & \text{☞} & A + \log 0 = -\infty \\
(\ [A]\ [\quad\quad\ ]\ ) & \text{☞} & A \times 0 = 0 \\
([[(A)]<[(<[\ ]>)]>) & \text{☞} & \#^A \div \#^\infty = 0 \\
(([[A]]\ [\quad\quad\ ]\ )) & \text{☞} & A^0 = 1
\end{array}$$

Chapter 10

The outer round-bracket in Dominion is not a necessity. An alternative form emphasizes the dominance of the square-bracket.[8]

dominion
$$(A\ [\ ]) = [\ ]$$

Although of fundamental importance, Dominion is a *theorem* arising from the three axioms that themselves give no hint of transgressing into non-numeric interpretations. Non-numeric forms are a structural consequence of exponential processes. Here is a purely structural demonstration of Dominion. I'll use the cardinality frame that we have been interpreting as multiplication.

```
                    ([A][ ])
create              ([A][ ])    A              <A>
mark                ([A][ ]) ([A][o])          <A>
collect                      ([A][o])          <A>
unmark                          A              <A>
cancel                         void
```

Void-equivalent forms are blind to multiplicity, so we also have as a corollary,

$$([A][\ ])\ ([A][\ ]) = ([A][\ ])$$

Promotion

We will need one more angle-bracket theorem. The Promotion theorem is essential, since it permits movement of angle-brackets across frames. The generic form of the theorem is

promotion
demote ⇄ promote
$$(A\ [<B>]) = <(A\ [B])>$$

Intervening angle-brackets do not interfere with promotion, although any other boundary type will. An alternate form of Promotion is

promotion
$$(A\ <[<B>]>) = <(A\ <[B]>)>$$

Reflection

Here is a demonstration of the generic form of Promotion.

```
(A [<B>])
(A [<B>]) (A [B]) <(A [B])>              create
(A [<B> B]) <(A [B])>                    collect
(A [     ]) <(A [B])>                    cancel
            <(A [B])>                    absorb
```

Demonstration of the alternate form of Promotion requires a different strategy. This demonstration is particularly enlightening, since it shows us how two distinct types of inversion in the interpretation, negation and division, interact. This is, incidentally, the longest demonstration in the volume.

```
 (A <     [<B>]>                          )
 (A <     [<B>]><[<B>]    > [<B>] )       create
<(A <     [<B>]><[<B>]    > [ B ] )>      promote
<(A <     [<B>]  [<B>]    > [ B ] )>      join
<(A <[ ([<B>]    [<B>]) ]> [ B ] )>       enfold
<(A <[<<([ B ]   [ B ])>>]> [ B ] )>      promote
<(A <[ ([ B ]    [ B ]) ]> [ B ] )>       unwrap
<(A <     [ B ]  [ B ]    > [ B ] )>      clarify
<(A <     [ B ]><[ B ]    > [ B ] )>      split
<(A <     [ B ]>                   )>     cancel
```

We call upon Reflection to generate a copy of <[<B>]> which can be promoted with its twin to lift the inner angle-bracket out of both. The pair of angle-brackets are then deleted via Involution. The remnant <[B]> is in the same container as [B], both cancel, leaving the desired result.

Promotion has an important structural limitation. As a theorem about frames, Promotion does not apply when square and round-brackets are exchanged. Visually,

$$[A (<B>)] \neq <[A (B)]>$$

Another structural constraint is that the promoted angle-bracket must enclose the *entire contents* of the inner square-brackets. That is, the transformation is on frames, not on selected contents. Let's illustrate this. Consider first two bracketed forms within an frame.

	(A [<B C>])	
create	(A [<B C>])	(A [B C]) <(A [B C])>
collect	(A [<B C> B C])	<(A [B C])>
cancel	(A [ ])	<(A [B C])>
absorb		<(A [B C])>

This works as expected, but now let's apply Promotion, with a slight variation.

	(A [<B C>])
split	(A [<B><C>])
ERROR	<<(A [B C])>>
unwrap	(A [B C])

Neither the individual <B> form nor the individual <C> form can be promoted due to the presence of the other.

$$(A\ [<B\ C>]) = <(A\ [B\ C])>$$
but
$$(A\ [<B>\ C]) \neq <(A\ [B\ C])>$$

Permeability

In general, boundaries possess the property of **permeability**. Any boundary is either transparent, semi-transparent, or opaque both to the forms it contains and to the forms in its context. Permeability specifies how, if at all, forms can cross a boundary without changing intention. Numeric forms are **opaque**.[9] Specifically, none of the James

opaque   *semipermeable*   *transparent*

transformation patterns permit a form to move across a single boundary. Opaqueness is the defining characteristic of numeric forms, the essence of Accumulation,

$$( )( ) \ne ( ) \ne (( ))$$

Semipermeability is the primary boundary property that distinguishes numeric forms from logic forms. Loosely the conventional name for the semi-transparency of logic boundaries is "implication".[10] **Transparent** boundaries are, of course, not boundaries.

Nested combinations of numeric boundaries can exhibit non-degenerate types of transparency. The Promotion theorem, for example, asserts that frames are transparent to an angle-bracket that contains the framed form.

$$(A\ [<B>]) = <(A\ [B])> \qquad \textbf{promotion}$$

Another example of transparency is the Inversion axiom.

$$([A]) = ([\ ])\ A = A$$
$$[(A)] = [(\ )]\ A = A \qquad \textbf{inversion}$$

The double boundaries ([...]) and [(...)] are transparent.[11]

We'll see another predominant form of three boundary transparency in Volume III. J-transparency is the premier theorem of transparency within the imaginary numeric realm. The specific configuration of three boundaries [<( )>] is transparent to its contents.

$$[<(A)>] = [<(\ )>]\ A \qquad \textbf{J-transparency}$$

10.2 Context

Angle-brackets reflect their contents. Depending on content and context, the angle-bracket can be interpreted as distinguishing positive from negative, addition from subtraction, multiplication from division, power from root, or power from logarithm.

Chapter 10

operation	expression	☞	form		embedded base
addition	A + B		A	B	
subtraction	A – B		A	<B>	
multiplication	A × B		([A]	[B])	([A] ( ))
division	A ÷ B		([A]	<[B]>)	([A]<( )>)
power	B^A		(([[B]]	[A]))	(A)
log_B	log_B A		(<[[B]]>[[A]])		[A]
root	B^(1/A)		(([[B]]	<[A]>))	((<[A]>))

Figure 10-1: *Arithmetic operations and their inverses*

Importantly, the multiple interpretations of the angle-bracket are imposed during transcription, while translating conventional notation into James forms and while reading James forms as conventional numeric expressions.

Figure 10-1 identifies the James form of each of the common algebraic operations, arranged to emphasize that an operation is converted into an inverse by including one new angle-bracket. The exception is the logarithm, which modifies boundary structure as well as adding an angle-bracket.

Embedded Base

Each boundary potentially implements a unary process. Crossing a boundary changes a value by applying the process. The permeability of a boundary determines which forms are permitted to cross without change.

The primary example of a **functional boundary** is the concept of a base. The positional rule for place-value numerals *"moving to the left multiplies by 10"* is the boundary rule *"crossing outward multiplies by 10"*. Both of these rules go by the name **base-10**.

Reflection

operation	equation	☞	form
subtraction	0 − 0 = 0		< > = *void*
	1 − 1 = 0		o < o > = *void*
	A − A = 0		A < A > = *void*
division	A ÷ A = 1		([A] < [A] >) = ( )
logarithm	log$_A$ A = 1		([[A]]<[[A]]>) = ( )
root does not reduce	A$^{1/A}$		(([[A]]< [A] >))

Figure 10-2: *Reflection within arithmetic operations*

The final column in Figure 10-1 shows the form of each operation when the base is implicit, or embedded into the boundary. Structurally, the form of the implicit base is B = (( )). This form applies to multiplication as well as to exponents.

Embedded bases expose the form of decimals.

$$A \div base \quad ☞ \quad ([A] <[(( ))]>)$$
$$([A] < ( ) >) \quad \textbf{decimal frame}$$

For example,

$$.5 \quad ☞ \quad ([5] <o>)$$

Another family of structure within angle-bracket forms is exposed by the void-based assertion of identity.

$$A = A \quad ☞ \quad A <A> = void$$

Each interpreted operation in Figure 10-2 can be directly reduced by Reflection. Each *reflected form* in Figure 10-2 creates an asymmetry. The Reflection axiom defines the behavior of that asymmetry. When we add, for example, we do not distinguish between the forms being added. However, when we subtract, one form is treated differently, the one that is subtracted. We can also say that a negative number (the different one) is added. In the case

Chapter 10

of multiplication, a reflected form is the one that divides. We can say that a unit reciprocal (the different one) is multiplied. In the case of powers, an inverted form is the one that provides the logarithm. The relationship in this case, though, is more complex since we must shift between logarithmic and exponential contexts during reflection.

One form, the root $^A\sqrt{A}$, does not incorporate Reflection. The Ath root is not the inverse of A, rather, it is the inverse of a base raised to the Ath power.

$$\#^{-A} \quad \☞ \quad (\text{<A>}) \qquad \#^{1/A} \quad \☞ \quad ((\text{<[A]>}))$$

Inverse Operations

outside
A <A> = *void*

inside
([A]) = A

Two of the James axioms assert a void-equivalence. Structural cancellation can occur both on the outside and on the inside of a boundary. Inversion cancels simply nested forms; the canceled forms share a context/content relation with one another. Reflection cancels two forms within the same container; the canceled forms share the same context relative to each other.

Objects and Functions

There are two different conventional conceptualizations of an inverse, the *inverse object* and the *inverse of a function*. In James algebra, object and function are confounded, the distinction between an inverse object and an inverse function is lost. An arbitrary numeric form A can be interpreted as an elementary object, for example o, or as a function, for example ( ).[12]

A + 0 = A
A + −A = 0

For an algebraic system (an operation and a set of objects) to be called a **group**, each object needs a matching inverse object. When the binary group operation acts upon an object and its inverse, the result is the **identity**. The identity for the operation + is 0, since positive and negative

250

numbers of the same magnitude add to 0. Since James algebra has no zero, addition as fusion has no identity.

It is best not to compare James algebra with group theory because the two approaches have so little in common.[13] Group theory classifies mathematical structures. A basic property of algebraic structure is associativity. The power function that serves as our primary interpretation of container structures is not associative.

$$A^{(B^C)} \neq (A^B)^C$$

The other way to think about inverses is an **inverse function** that cancels the effect of a given function to yield no change. We can add 2 to 5, and by subtracting 2 from the result, we return to the original 5. Subtraction is the inverse function for addition. The inverse *object* leads to the identity, while the inverse *function* leads back to the initial value prior to applying the function.

$$(5 + 2) - 2 = 5$$

The form of the Inversion axiom is analogous to a functional inverse. The round container is the inverse of the square container. When each is applied to the other, they cancel, leaving us with the original form A.

$$([A]) = [(A)] = A \qquad \textbf{inversion}$$

The Reflection axiom is analogous to an object combined with an inverse object. There is no "operation" that binds the two forms A and <A> together. Reflection is composed of two contains relations, the outer-container contains A and the same outer container contains <A>. It is these two containment relations that cancel.

$$(A \ \text{<A>}) = (\ ) \qquad \textbf{reflection}$$

Figure 10-3 shows the non-reduced James forms of the conventional inverses. The inverse object in the interpretation is marked I. Conventional addition and multiplication have object-like inverses, while power and logarithm have function-like inverses. Roots as inverses exhibit a quite different boundary structure.

251

Chapter 10

form	☞	operation	inverse	I	value
A <A>		A + −A	A + I	−A	0
([A][(<[A]>)])		A x 1/A	A x I	1/A	1
([A])		#log# A	#I	log#A	A
[(A)]		log# #A	log# I	#A	A
(([(<[A]>)][[(A)]]))		(# A)1/A	(# A)I	1/A	#

Figure 10-3: *The containment structure of inverse operations*

Reflected Forms

Figure 10-4 shows the structure of the reflection pairs for each of the four common operations of arithmetic, and provides a summary of this section. Inversion pairs that `clarify` are emphasized. From this perspective it is easy to see the structural similarity of inverse operations. With the exception of addition, each requires `clarify` to untangle the components of the reflection pair.

For **add/subtract**, the behavior of the reflection <A> is defined by the Reflection axiom, which also provides the definition of the angle-bracket < >. The James forms in Figure 10-3 show that + has an object-like inverse, yielding the identity 0.

For **multiply/divide**, the James form is a combination of function-like and object-like structure. The inverse object is the *unit fraction*.

$$1/A \ \text{☞} \ (<[A]>)$$

The product of A and its reciprocal 1/A reduces to the multiplicative identity 1. However, the form of the reciprocal includes an inversion pair separated by < >, endowing it with operational properties. Reducing the form of the product yields a reflection pair, the object-like inverse structure.

Reflection

operation	form	reflect	interpret
addition			A + −A
	A <A>	A <A>	
multiplication			A × 1/A
	([A][(<[A]>)])		
	([A] <[A]>)	[A]<[A]>	
power			$B^{\log_B A}$
	((([B]][(<[[B]]>[[A]])]))		
	(([[B]] <[[B]]>[[A]]))	<[[B]]>[[B]]	
logarithm			$\log_B B^A$
	(<[[B]]>[[(([[B]][A]))]])		
	(<[[B]]> [[B]][A])	<[[B]]>[[B]]	

Figure 10-4: *The reflection structure of inverse operations*

```
A × 1/A  ☞  ([A][ (<[A]>) ])
            ([A]    <[A]>    )        clarify
            (                )  ☞  1  cancel
```

The reflection pair [A] and <[A]> is nested within an outer round-bracket. After canceling, the round-bracket context remains as a residual that we interpret as 1.[14] Multiplication has a different identity than addition because the inverse of what we call division is an exponential power relation. The James form (<[A]>) can be read as a negative power, $\#^{-\log_\# A}$.

Multiplication is *addition* of powers. In conventional notation

$$A \times 1/A = A^1 \times A^{-1} = A^{1+(-1)} = A^0$$

Were it not the case that anything to the zero power is defined to be 1, the identity for multiplication would not be 1. A^0 must equal 1 for the multiplicative identity, $1 \times A = A$, to work.

Chapter 10

For **power/logarithm**, the James inverse is function-like. Using embedded bases, (A) is a power and [A] is the corresponding logarithm. The inverse relation is defined by the Inversion axiom, and is structurally independent of the base.

$$\log_\# \#^A = A \quad ☞ \quad [(A)] = A$$
$$\#^{\log_\# A} = A \quad ☞ \quad ([A]) = A$$

Alternatively, we can use the form of power and the form of logarithm with a *specific* base.

$$B^A \quad ☞ \quad ((\; [[B]] \; [A] \;))$$
$$\log_B A \quad ☞ \quad (<[[B]]>[[A]])$$

The form of the power/logarithm function inverse is fundamentally different since the nesting levels in the logarithm are not balanced. The logarithm is the only conventional function with the property of unbalanced frame structure. Using the specific base clearly generates a function-like inverse. The reflected form must replace A, the variable standing in place of the specific power. Like multiplication, reducing this resulting form first requires Inversion to expose the Reflection pair, a characteristic of all inverses expressed in conventional notation.

```
             B^(log_B A)  ☞  (([[B]][ (<[[B]]>[[A]]) ]))
   clarify                   (([[B]]    <[[B]]>[[A]]    ))
   cancel                    ((             [[A]]       ))
   clarify                                    A


             log_B B^A    ☞  (<[[B]]>[[ (([[B]][A])) ]])
   clarify                   (<[[B]]>      [[B]][A]      )
   cancel                    (                [A]        )
   clarify                                    A
```

It is the base rather than the power that participates in Reflection. From these power/logarithm structures we can see why James computations are base-free. In each

254

case, the base reduces to *void*. Whether the base is represented as B or as # does not matter.

For **power/root**, the inverse structure is that of multiplication, but at the level of exponents. Multiplication/division can be seen as addition of exponents, similarly power/root can be seen as multiplication of exponents. The power/logarithm inverse relationship applies Reflection to the *base* of each. The power/root pseudo-inverse relation applies Reflection to the *power* of each. Again we can examine the James forms.

$\#^A$ ☞ `(( [[#]] [A] )) = ( A )`

$\#^{1/A}$ ☞ `(( [[#]] <[A]>)) = ((<[A]>))`

The function-like inverse relationship is exposed by nesting one form of inverse within the other. Rather than replacing the power as we did with power/logarithm, we replace the base. Since it is the base that participates in the power/logarithm inverse, this change should leave the base untouched, while canceling the exponents. The base itself is arbitrary, while any particular base creates a reflection pair.

$(\#^A)^{1/A}$ ☞ `(([[(A)]] <[A]>))`
           `(([  A  ] <[A]>))`   clarify
           `((          )) = #`   cancel

$(\#^{1/A})^A$ ☞ `(([[((<[A]>))]] [A]))`
             `((   <[A]>     [A]))`   clarify
             `((              )) = #`   cancel

We can see that the reflection pair [A]<[A]> is identical to that of multiplication, which means that power/root does *not* identify a different type of inverse relationship. We might say that the *object inverse* of a power is a logarithm, while the *operator inverse* of a power is a root.

When the conventional order of application is reversed, the location of the [A] and <[A]> forms is exchanged. Since James forms have no concept of ordering within a container, this example illustrates that our conventional

Chapter 10

expression	☞	James forms	simplified
		COUNTING	
2 − 1		oo < o >	o
1 − 1		o < o >	void
0 − 1		< o >	
		SUBTRACTION	
0 − B		< B >	
A − B		A < B >	
A + −B		A < B >	
		DIVISION	
2 / 1		([2] <[o]>)	2
1 / 1		([o] <[o]>)	o
1 / 2		([o] <[2]>)	(<[2]>)
1 / B		([o] <[B]>)	(<[B]>)
A / B		([A] <[B]>)	
		LOGARITHM	
$\log_\# 1$		(<[[#]]>[[o]])	void
$\log_\# \#$		(<[[#]]>[[#]])	o
$\log_\# A$		(<[[#]]>[[A]])	[A]
$\log_B A$		(<[[B]]>[[A]])	
		ROOT	
B^{-A}		((([B]] [<A>]))	
$B^{1/1}$		((([B]]<[o]>))	B
$B^{1/2}$		((([B]]<[2]>))	
$B^{1/A}$		((([B]]<[A]>))	
$\#^{-A}$		((([#]] [<A>]))	(< A >)
$\#^{1/2}$		((([#]] [<2>]))	(<[2]>)
$\#^{1/A}$		((([#]]<[A]>))	(((<[A]>))
		IMAGINARY	
$\log_\# -1$		[<o>]	J

Figure 10-5: *Angle-bracket forms*

notation for composing exponential operations reduces to a choice of which reflected form to enfold.

The computed inverses exhibit remarkable similarity. The shorter transformation sequences call upon the Reflection axiom, while the longer sequences require clarifying some inversion pairs first, before calling upon Reflection (followed by more clarifying). What differs is the structure of the forms that trigger canceling. The entire collection of inverses require only two axioms, Inversion and Reflection. This suggests a hidden simplicity underneath the conventional concepts of inverse objects and functions.

Figure 10-5 reiterates the shared structure of the inverse algebraic operations, this time providing examples of diverse structural patterns from both arithmetic and algebra.

10.3 Remarks

In Volume III we'll explore the rather surprising interpretation of the square-bracket, [ ], as *negative* infinity. Polarity made an early, if somewhat obscured, appearance in Chapter 7, prior to the introduction of the Reflection axiom. What if [ ] were positive? One way to do this is to build the *concept* of an inverse operation directly into the functionality of the square-bracket. This new container could be represented as a double-struck square-bracket and interpreted as +∞.

$$\llbracket A \rrbracket =def= [<A>]$$

James algebra can now be fully expressed using only two types of bracket, ( ) and ⟦ ⟧. The conventional operations of arithmetic no longer have polar duality as an essential property; inverse operations become contextual. This dialect is explored in Chapter 20 of Volume II.

But now we'll apply the axiomatic basis of James algebra to help us to explore the structure of common numbers.

Chapter 10

Endnotes

1. **opening quote:** P. Lockhart (2012) *Measurement* p.204.

2. **<A> is its negative inverse:** The approach of unit-ensembles, in contrast, is to construct a unique type of negative unit, ◊, rather than to attach an operator to the positive unit as we do with natural numbers, i.e. −N, and as we will do here with a reflecting boundary, <N>. We usually interpret a negative number as a modified positive number, however the James reflecting boundary is not bound to its specific contents. The operations associated with any container are independent of the contents of that container.

3. **expressed by the fundamental operation of boundary reflection:** It's tempting to stretch this metaphor a bit.

    < >   *reflection*
    ( )   *translation*
    [ ]   *rotation*

    Strictly as an analogy, translation can be represented by the accumulation of round-brackets. Rotation can be represented by the introduction of the complex-plane as an interpretation of square brackets. More on this in Volume III.

4. **that arises directly from Reflection:** This theorem has the same structure as Merge in Chapter 2, indicating a deeper communality between merging groups and negation.

5. **The Reaction theorem exchanges the reflection boundary across forms:** The void-equivalent form of Reaction,

    <<A> B> <A <B>> = *void*

    has the same boundary structure as logical XOR, logical NOT EQUAL.

6. **angle-brackets are needed now but will be shown to be superfluous later:** Volume II introduces a composite boundary, ⟦A⟧ = [<A>], that integrates the angle bracket with the square bracket.

7. **A can be any numeric form:** Although we include multiplication by 0 in the conventional multiplication tables for whole numbers, it is an entirely different process. It is a conceptual confusion to say "0 times 5 is 0" because "times" is not the process that converts five into nothing.

8. **the dominance of the square-bracket:** Still we must take care to acknowledge that [ ] does not interact with numeric forms in the same container. The square-bracket, technically, does not dominate or degenerate. It merely fills, completely, any container it may be in.

Jeffrey James calls [ ] the *black hole*, since it appears to absorb everything it shares a container with. This makes [ ] rather like the Borg in StarTrek. Although a singular entity, the Borg refers to itself as *we*. "We are the Borg. You will be assimilated. Resistance is futile." We'll explore assimilation in the form of indeterminate variables in Volume III.

9. **Numeric forms are opaque:** After reduction, forms that contain an empty square-bracket, [ ], are not numeric. Similarly, void-equivalent forms are not numeric, since *void* is not a numeric value.

10. **semi-transparency of logic boundaries is "implication":** *Logical boundaries* are semipermeable. This in fact is the primary difference between numeric and logic systems. Although there are two kinds of numeric inequality (greater-than and less-than), one does not leak into the other. The numeric boundary is impermeable. Logic, in contrast, is based on inference, which is one directional. A IMPLIES B does not mean that B IMPLIES A. In boundary logic, semipermeability is called Pervasion and it is the defining structure of boundary logic. The outside of a logical boundary pervades the inside. If it is in the context then it is arbitrarily also in the content. Below, ⟨ ⟩ is the semipermeable logic boundary.

$$A \langle A\ B \rangle = A \langle B \rangle \qquad \textbf{pervasion}$$

The structural similarities and differences between logic and numerics are displayed in Figure 15-1.

11. **double boundaries ([...]) and [(...)] are completely transparent:** Void-equivalent inversion pairs are the primary detritus that leads to the apparent diversity of conventional expressions.

12. **or as a function, for example ( ):** The simple process that constructs the inverse of any James form is described in Volume II.

13. **the two approaches have so little in common:** The same can be said of conventional functions compared to conventional relations. The two worlds have different evolutions, different mathematical perspectives, and different

structural rules well beyond the existence and uniqueness mapping constraints of functions. Function theory grew hand-in-hand with calculus several hundreds of years ago, while becoming dependent upon the concept of continuity. Relational theory grew with computational databases less that fifty years ago within a discrete digital context. Generally, functions are the domain of mathematics while relations are the domain of computer science.

14. **remains as a residual that we interpret as 1:** Group theory fails to identify why its structures exist. *Zero* is the additive null-object because addition-as-fusion occurs in the value-free context of shell-brackets. That is, addition is a boundary process that has nothing (i.e. zero) to do with numeric value. *One* is the multiplicative null-object because it occurs within a value-laden context of the numeric unit, ( ). Simple multiplication is contextual addition, so its null-object must exist to provide a numeric context. The two group-theoretic null-objects simply refer to the presence or absence of a numeric container.

Chapter 11

Numbers

...numbers have neither substance, nor meaning, nor qualities. They are nothing but marks, and all that is in them we have put into them by the simple rule of straight succession.[1]
— *Hermann Weyl (1959)*

We have explored sufficient James structure now to describe the range of conventional numbers. Natural numbers themselves are easy, they are the result of the accumulation of tallies. It is first our conventional notation, the vaunted place-value system; second our operations, the revered addition and multiplication algorithms; and finally our real numbers that lead us into the mire that every grade school student has endured. On the way to an organic understanding of arithmetic, in this chapter we'll deconstruct the conventional number system, eventually arriving at the physical and conceptual inaccessibility of real numbers, and at the embarrassing confusion of calling the numbers that are most divorced from reality, "real".

Babylonian numerals circa 2000 BCE

11.1 Numerals

Our Hindu-Arabic number system has evolved from the tallies and grouping of Babylonia over 4,000 years ago, and from the conventions of Indian and Persian notations over 2,000 years ago. Five inventions converge to give us the way we now write the natural numbers:

— uniform grouping (Mesopotamia, c.2000 BCE)
— positional notation (Babylonia, c. 2000 BCE)
— digits   (India/Persia, c. 600 CE)
— zero   (India, c. 600 CE)
— place-value (India, c. 800 CE)

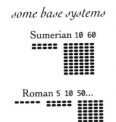

some base systems

Sumerian 10 60

Roman 5 10 50...

Mayan 20

Arabic 10

Digits are tightly connected to the base system. When a count reaches a certain threshold, i.e. the **base**, notation reverts to reuse of the same digits. In base-10 we have nine digits and zero. We reuse the ten symbols in different ordered strings to represent different magnitudes. Uniform grouping immediately calls for putting each power of the base in a different location, thus positional notation. Zero is needed to keep positions unambiguous whenever no digit is needed. The important idea is that place-value is shorthand for a rather complex construction, the *polynomial number*.

Historian Otto Neugebauer comments that place-value has been hailed as "undoubtedly one of the most fertile inventions of humanity."[2] Much of the system, along with its operations, penetrated Western culture in the sixteenth century. Negative numbers continued to meet resistance well into the nineteenth century. Our notation for large numbers is simply an abbreviation that improves upon ancient methods of computation but in itself still carries the unnecessary burdens of linearity, polynomiality, and fabrication.

The boundary equivalent to place-value is **depth-value**, which uses successive nesting to represent powers. Depth-value arithmetic is described in Chapter 3. We are not using depth-value within James numbers, although it is available as an abbreviation. By replacing position by depth we eliminate the sequential dependencies of typography and gain the benefits of spatial parallelism. We eliminate an additively fragmented polynomial structure in favor of a unified factored form. And the fabricated notion of 0 returns to non-existence.

Numbers

number type	standard expression ☞	James form
zero	0	
natural number	3	ooo
negative number	−3	<3>
unit fraction	1/3	(<[3]>)
rational	3/4	([3]<[4]>)
other types of conventional numbers		
mixed number	2 3/4	2 ([3]<[4]>)
prime factors	$3^2 \times 5$	([5]([[3]][2]))
explicit base-2	1101	([2 o][2][2]) o
implicit base-2	1101	(((o) o)) o
irrational	$\sqrt{3}$	(([[3]]<[2]>))
e, *transcendental* base-e	e	(o)
imaginary and non-numeric forms		
J, *James imaginary*	log−1	[<o>]
i, *imaginary*	$\sqrt{-1}$	(([J]<[2]>))
complex	4+3i	4 ([3]([J]<[2]>))
pi, *transcendental*	π	([J]([J]<[2]>))
negative infinity	−∞	[ ]
positive infinity	∞	<[ ]>

Figure 11-1: *Types of numbers*

Variety

Ordinary numbers are extraordinarily diverse. One objective of James numbers is to provide some organization to this diversity. We have become inured to the types and representations of our numbers. Figure 11-1 presents some of this diversity and the corresponding James forms. Rather than diversity by definition and by notation, the James approach is to display diversity as variation in the structure of types of nested containers.

Figure 11-1 incorporates common digits within the James forms to abbreviate explicit ensembles of tallies. Symbolic condensation is optional, and very handy when we're working with containment patterns. The primary feature of Figure 11-1 is that the diversity of common number types reduces to structural variations of nested containers.

Our common numbers are encoded. To know what the radical sign √ means, we need to look it up. All of the codes eventually reduce to addition and multiplication. In Figure 11-1 we see at least ten encoded conventional operations (-, /, space, power, x, place-value, √, log, +, labels). The James notation too is encoded. Given that we understand that space is just containment, there are three operations, the three boundary types.[3]

A primary division between types of number is **cardinal** or **ordinal**. Cardinal numbers begin with a unit and then repeatedly add that unit until a desired magnitude is reached. The common digits 0 through 9 stand in place of cardinal numbers. Ordinal numbers (*first, second, third,*...) use a different symbol for each new number, in effect forcing an ordering upon names instead of units. Their ordering is determined by the number of repetitions of addition, but not by the magnitude of what is added.

While Figure 11-1 shows the various types of numbers, Figure 11-2 shows several *varieties of representation* we can use for the same number. The base-10 examples show six ways that we can write 654 in standard notations. Our conventional numbers have place-value, additive, polynomial and factored formats. Decimal and scientific notation explicitly identify the position of the units column, effectively combining numbers greater than 1 with those less than 1. Decimal form, for example, has the structure of multiplying by the unit fraction of the base.

```
                    .4    ☞  ([4]      <[b]>   )
     enfold                  ([4][ (<[b]>) ])   ☞  4 x 1/b
```

Numbers

notation	expression	☞	form
BASE-10			
place-value	654		⦅⦅(6)⦆5⦆4
additive	600 + 50 + 4		([6]2)([5]o) 4
decimal	.654		([6]<o>)([5]<2>)([4]<3>)
polynomial	$(6 \times 10^2)+(5 \times 10^1)+(4 \times 10^0)$		([6]2)([5]o) 4
factored	$((6 \times 10) + 5) \times 10) + 4$		([([6]o)5]o)4
scientific	6.54×10^2		([6([5([4]<o>)]<o>)] 2)
BASE-B			
→ *place-value*	$654_{base\text{-}b}$		⦅⦅(6)⦆5⦆4
→ *additive*	$(600 + 50 + 4)_{base\text{-}b}$		([6]([[b]][2]))([5][b]) 4
→ *decimal*	$.654_{base\text{-}b}$		([6([5([4]<[b]>)]<[b]>)]<[b]>)
polynomial	$(6 \times b^2)+(5 \times b^1)+(4 \times b^0)$		([6]([[b]][2]))([5][b]) 4
factored	$((6 \times b) + 5) \times b) + 4$		([([6][b])5][b])4

Figure 11-2: *Varieties of numeric encoding*

The depth-value form that corresponds to place-value in Figure 11-2 uses a **double-struck round-bracket** ⦅N⦆ as an abbreviation for the unit magnitude frame, ([N] o).

The three examples marked by an arrow, →, each incorporate an implicit base. We need to know the base in order to know what these numbers mean. Since the base is hidden, the conventional forms for place-value, additive, and decimal numbers look the same whether we are in base-10 or base-b. We will use *embedded* bases later to both simplify the notation for numbers and to add parallelism.

The explicit base-b entries in Figure 11-2 grow large in any notation since the base must be recorded visually. The **Principle of Communality** states that when a property belongs to all members of a collection, then it is a property

Chapter 11

of the collection rather than the members. Thus the multiple reference to the base for each digit of a number is justified only by convention, a residual of thinking of numbers as sequential polynomial expressions.

Polynomials

Athenian numerals circa 500 BCE

Babylonian, Egyptian and early Greek numerals were **additive**, a sum consists of the juxtaposition of individual symbols. The Athenian number system, circa 500 BCE, was *acrophonic*: numerals were the letters of the alphabet, with the ordinal position of letters determining the cardinal value that the letter represented. Special letters were reserved for **5, 10, 100, 1000,** and **10000**. Instead of grouping within a uniform base system, number groups were specialized for specific magnitudes. The Athenians also had a multiplicative principle. They combined place-value with depth-value. To multiply **5** by a base, the base is placed *inside* the letter/number for **5**.[4]

numbers are base-10 polynomials

The key structural idea underlying modern numbers is to express magnitude as an **algebraic polynomial**. To deconstruct place-value shorthand, write each *digit* in scientific notation, with the order of magnitude made explicit.

$$5683 = 5000 + 200 + 80 + 3$$
$$= (5 \times 10^3) + (6 \times 10^2) + (8 \times 10^1) + (3 \times 10^0)$$
☞ ([5] 3) ([6] 2) ([8] 0) 3

The parts of a polynomial add up, so we might consider our universal notation for whole numbers to be an additive system complemented by uniform bundles of **10**. The corresponding James form is also shown. It consists of four independent components, each keeps track of a digit and its order of magnitude, expressed as a *magnitude frame*. I've elected to put the digit (which is nested within a square-bracket) on the left and the power on the right, although their relative positions are arbitrary. The James form is quite natural to read, it is similar to scientific notation. Here is the general form and an example.

```
([A] N)    ☞    A x 10^N
([5] 3)    ☞    5000
```

We are calling upon a new type of frame, the **magnitude frame**, which is an interpretation of the generic form of a frame.

(frame-type [frame-form])
(magnitude [digit])   ☞   *digit* x 10^*magnitude*

A generic natural number in polynomial form looks like this:

$$a_n b^n + a_{n-1} b^{n-1} + a_{n-2} b^{n-2} + \ldots + a_1 b^1 + a_0 b^0$$

Each variable a_i stands in place of a coefficient (a digit in a particular position).[5] The corresponding generic James form, in hybrid notation, is below. It is stacked for readability.

```
([a_n    ] ([[b]][n       ]))
([a_n<o> ] ([[b]][n<o>    ]))
([a_n<oo>] ([[b]][n<oo>   ]))
 . . .
([a_1    ] ([[b]][o       ]))
([a_0    ] ([[b]][        ]))
```

When the base is embedded within the round- and square-brackets, the James polynomial form is simpler:

```
([a_n    ] n      )
([a_n<o> ] n<o>   )
([a_n<oo>] n<oo>  )
 . . .
([a_1    ] o      )
([a_0    ]        )
```

The polynomial structure is that of a *sum* of *products* of *powers*. With all of the common arithmetic operations bundled together, no wonder it is so expressive. Lakoff and Núñez consider polynomials to be nearly organic:

Chapter 11

Our linear, positional, polynomial-based notational system is an optimal solution to the constraint placed upon us by our bodies (our arms and our gaze), our cognitive limitations (visual perception and attention, memory, parsing ability), and possibilities given by conceptual metaphor.[6]

Alternative opinions suggest that linear and positional notations reflect the limitations of historical technological display media, that polynomials have more to do with culture than with physiology, and that computational algorithms are artifacts of devices rather than deliberations.

Factors

There is a different canonical way to express a conventional number, the **maximally factored expression**. Polynomial notation adds the magnitudes together. Numbers can also be written as successive embedded multiplications. This equivalent factored notation is rarely used, and we do not have an efficient conventional notation. Here's what the example 5683 looks like.

$$5683 = (((((5 \times 10) + 6) \times 10) + 8) \times 10) + 3$$

☞ ([([([5] o) 6] o) 8] o) 3

The factored form is more natural than the polynomial form for James integers, since we have been interpreting the content of round-boundaries as raised to a power.

It is also useful to compare the two James versions, the additive and the multiplicative:

polynomial	5683 ☞	([5] 3)([6] 2)([8] o) 3
factored	☞	([([([5] o) 6] o) 8] o) 3

The polynomial form is *counting* exponents from left-to-right, the multiplicative form is *tallying* exponents. The polynomial format emphasizes additive components at the cost of specifying each individual order of magnitude. The multiplicative form nests uniform **unit magnitude frames**.

$$([N] \quad o \quad) \qquad \textit{unit magnitude frame}$$
$$([N][(o)]) \quad ☞ \quad N \times \textit{base} \qquad \text{enfold}$$

Let's use base-10. We can elect to absorb the "x10" into the outer round-bracket. By absorbing *the unit magnitude frame* into the round bracket, we change it's definition from identifying an arbitrary exponential base to identifying a specific base.

$$5683 \quad ☞ \quad ([\ ([\ ([5] \ o) \quad 6] \ o) \quad 8] \ o) \quad 3$$
$$☞ \quad (\!(\ (\!(\ (\!(5 \ )\!) \quad 6 \ )\!) \quad 8 \ )\!) \quad 3 \qquad \text{substitute}$$

This redefinition of the parens round boundary that was introduced in Chapter 3 yields the **depth-value double-struck round-brackets**.

$$(\!(N)\!) \ =\!\textit{def}\!= \ ([N] \ o) \qquad \textbf{depth-value}$$

Both the polynomial and the factored James forms provide condensed notations for large ensembles of single units. Neither of these notations is shorthand, since each displays the order of magnitude directly rather than requiring memorized rules such as those associated with adding and multiplying positional numbers.

Counting the order of magnitude, counting "how many zeros are behind the number", is literally the source of complexity in grade school multiplication and division algorithms. Place-value notation imposes a heavy cost upon its computational algorithms. Place-value is designed for reading numbers, not for operating upon numbers. In contrast, tallying order of magnitude while storing the "number of zeros" in descending *depth* permits homogeneous treatment of each nesting context through the use of the unit magnitude frame. Reading numbers is more difficult due to the explicit depth delimiters, while operating upon numbers becomes significantly easier.

Chapter 11

$$(((((((a_n \times b) + a_{n-1}) \times b) + a_{n-2}) \times b) + \ldots) + a_1) \times b) + a_0$$

☞ $([([([([a_n][b]) \; a_{n-1}] \; [b]) \; a_{n-2}] \; [b]) \; \ldots \quad a_1] \; [b]) \; a_0$

$([([([([a_n] \; \circ \;) \; a_{n-1}] \quad \circ \;) \; a_{n-2}] \quad \circ \;) \ldots \quad a_1] \quad \circ \;) \; a_0$

$(\!(\!(\!(\!(\; a_n \qquad)\; a_{n-1} \qquad)\!)\; a_{n-2} \qquad)\!) \ldots \quad a_1 \qquad)\!)\; a_0$

Figure 11-3: *Factored number to depth-value*

Depth-value encoding also creates a structure that is parallel rather than sequential.[7] Of course the descriptors "additive" and "multiplicative" are projections from our interpretation. More accurately we might identify the James forms as disperse and collect, respectively. Arrangement converts between these forms.

$\qquad\qquad\qquad ([5] \; ooo) \; ([6] \; oo) \; ([8] \; o) \qquad 3$
collect $\qquad ([([([5] \; \circ \;) \qquad 6] \; \circ \;) \quad 8] \; \circ \;) \qquad 3$

Figure 11-3 shows the generic factored form for conventional integers, followed by three transformations to successively simpler versions of the generic James form. The first James form, after the finger, is a literal transcription of the conventional polynomial expression. The second embeds the magnitude of the base into the meaning of the round- and square-brackets. The third is the double-struck abbreviation.

$[b] = [(())] = ()$

Figure 11-3 has a notational problem, since *ellipses* are intended to display sequence rather than depth of nesting. This and the general awkwardness of the conventional factored expression emphasize that conventional notation is not adapted for nested forms in specific and for embedded (iterated) operations in general. Our predisposition toward sequential numeric processes is built deeply into our common notation, a clear example in which *notation is not neutral* to the ideas it expresses.[8]

Common fractions are easily represented within the same format.

$$3/8 \; ☞ \; ([3]<[8]>)$$

Like conventional numbers, James decimals can be expressed as a continuation of a polynomial expression and as a factored form.

$$.375 = (3 \times 10^{-1}) + (7 \times 10^{-2}) + (5 \times 10^{-3})$$

☞ ([3] <o>) ([7] <2>) ([5] <3>)	*polynomial form*
([3 ([7 ([5] <o>)] <o>)] <o>)	*factored form*
([3 ([7 ([5] (J))] (J))] (J))	*(J)-frame form*

The (J)-frame version on the last line is a focus in Volume III. Within James algebra, negative numbers and negative exponents are both imaginary.

11.2 Real Numbers

There is a corresponding James form for every integer and for every rational number. Do James forms map to the real numbers? Mathematician Ian Stewart:

> The real numbers are one of the most audacious idealizations made by the human mind, but they were used happily for centuries before anybody worried about the logic behind them.[9]

The *real numbers* require us to leave the physical constraints of containers, since each of them has an infinity of digits. The reals extend the rationals by including the **irrational numbers**, which cluster into three broad categories: algebraic, transcendental, and lawless. The **algebraic reals** have corresponding constructions as solutions to polynomial equations, and they are countable. We can thus call upon algebraic functions that use the arithmetic operators $\{+, -, \times, \div, \wedge, \sqrt{}\}$ to build an equivalent James form. Perhaps the simplest example of an algebraic irrational is √2. Since there is a James form for fractional exponents, there is a James form for any number expressed with a rational exponent.

$$\sqrt{3} = 3^{1/2} \ \text{☞} \ (([[3]]<[2]>))$$

The **transcendental numbers** are real numbers that are not roots of polynomial expressions but do correspond to a defined symbolic function. The paramount example of a transcendental number is π. The number π cannot be expressed by any finite assembly of algebraic operations. The transcendentals themselves are not denumerable. Paul Lockhart observes:

> Transcendental numbers — and there are lots of them — are simply beyond the power of algebra to describe.[10]

There are an uncountable infinity of transcendental numbers, and James forms can represent only a countable number of them. These include the trigonometric, hyperbolic and logarithmic numbers. Since we have an interpretation of square-brackets as logarithms, we can construct logarithms of both rational and algebraic irrational numbers, as well as using any of them as the base of a logarithm. This then provides many examples of transcendental numbers that can be expressed as James forms. For example,

$$\log_B N \ \text{☞} \ ([[B]]<[N]>)$$
$$\pi \ \text{☞} \ ([[<o>]]([[<o>]]<[2]>))$$
$$2 \cos \theta \ \text{☞} \ (([\theta] \ J/2))(<([\theta] \ J/2)>)$$

The James forms of trigonometric functions implicate the structure of complex numbers and the mapping from exponential expressions to locations in the complex plane. They each contain the James imaginary, J. For example,

$J = 2 \log_\# i$

$$\pi \ \text{☞} \ ([[<o>]]([[<o>]]<[2]>))$$
substitute $\quad ([\ J \ ]([\ J \ ]<[2]>))$
hybrid $\quad\quad ([\ J \ ] \quad J/2 \quad)$

Here we have used forms that are themselves not rational (i.e. [2]) to describe the transcendental number π. This makes our interpretation of container-based forms inherently more expressive than algebraic expressions from the start. However, the physical semantics of James forms is not commensurable with real numbers defined solely by infinite series.

Although a James form can be constructed for specific real numbers that make sense in the context of defined expressions such as logarithms, roots, trigonometric functions and hyperbolic functions, there is no presumption that all real numbers (the uncountable infinity of them) have James forms.[11] The James forms are countable.

The Cut

Richard Dedekind formalized the real numbers in 1872. But Dedekind did not take the step into infinite complexity. "...we must endeavor completely to define irrational numbers by means of the rational numbers alone."[12] From here, Dedekind then steps backward into infinity.

Richard Dedekind
1831–1916

> I regard the whole of arithmetic as a necessary, or at least natural, consequence of the simplest arithmetic act, that of counting, and counting itself is nothing else than the successive creation of the infinite series of positive integers in which each individual is defined by the one immediately preceding; the simplest act is the passing from an already-formed individual to the consecutive new one to be formed.[13]

The idea of a **Dedekind cut** is to define the *spaces between* rational numbers as irrational. Real numbers are a boundary between two sets of rational numbers, one set all being smaller, the other set all being larger. It is as though we could identify an irrational such as π by bounding it from above and from below with successively closer

Chapter 11

rationals. Just like the Greeks did 2500 years earlier in approximating the value of π by successively nesting geometric shapes.

Wittgenstein found the now accepted definition of the real numbers to be nonsense. The Dedekind description does not provide an algorithm with which to construct or to enumerate the reals. Numbers are operations, they are rules as well as objects. We can make closer and closer approximations of an irrational such as √2; the notation for the irrational number itself *is* the algorithm to make these approximations. However, √2 does not exist *as an object* because for Wittgenstein no number is an object:

> In mathematics *everything* is algorithm and *nothing* is meaning; even when it doesn't look like that because we seem to be using *words* to talk *about* mathematical things. Even these words are used to construct an algorithm.[14]

Ludwig Wittgenstein
1889–1951

From the perspective of computer science, Wittgenstein's philosophy of mathematics is appealing. It is unnecessary, and perhaps unreasonable, for a finite system to be able to represent any arbitrary irrational number. Mathematical forms that are encoded by a finite binary sequence are constructible, indeed algorithmic. Each James form, for example, defines a process for generating a particular number. The form of an irrational number, for both boundary and conventional notations, is a computational specification, and computational specifications cannot call upon infinite processes.

To Wittgenstein, there is no reason or criterion "for the irrational numbers being complete".[15] Wittgenstein thus bans our third category of irrational numbers, the **lawless irrationals** that have no identifiable structure in their (infinite) sequence of decimals. James forms cannot express lawless irrationals, by definition no notation can. Some lawless irrational numbers are **chaotic**. We cannot know in advance which digit in a lawless number comes next since, being lawless, there is no mathematical process available to predict the structure. These numbers are *not* random, they are indeed deterministic. They are just immune to mathematical abstraction. There are lawless numbers that cannot be known or represented by any process, again not necessarily random since statistical randomness can be evaluated. We are fast approaching a central question about real numbers: are numbers that cannot be identified by any mathematical technique still legitimate as numbers?

Only a very special subgroup of real numbers can be specified, only those for which there is an algorithmic process that generates them to any specified degree of accuracy. The continuum of real numbers is not only imaginary, it is irretrievable by imagination. We are but a small step away from philosopher Mary Leng's modern perspective that *mathematics is a well-constructed fiction*.

> Mathematical hypotheses, on my view, are best thought of not as truths by convention (for they do not have the status of *truths*), but rather, as conventionally adopted useful fictions.[16]

Abstract mathematics has evolved into a somewhat self-indulgent dance. Like writing it is a communion between our thoughts inside and our symbolic squiggles outside. Brian Rotman:

> In other words, mathematics is a rigorous inscriptional fantasy: the insistence on writing determining what can be legitimately imagined, and the ongoing process of imagining controlling what mathematicians can meaningfully and usefully write down.[17]

Drawing icons and diagrams and containment boundaries places a different type of rigorous constraint upon mathematical structure, that of physical realizability. James forms thus identify the *constructible* real numbers.

Chapter 11

Computation

Historically, real numbers are riding upon the demands and the successes of calculus. Linearity and continuity require the continuum, the boundless infinity populated with an uncountable infinity of numbers each infinitely long. *Computation*, however, is offering a replacement for the numeric epistemology of calculus: discrete, digital, finite, embodied math. Rigorous theorists like Stephen Wolfram and John Wheeler have constructed finite structural models of computation, physics, and science. Here's Wheeler:

> The physical continuum, and with it all the beautiful machinery of physics, is myth, is idealization. Existence, what we call reality, is built on the discrete.[18]

An important perspective is that the only irrational numbers we know of are the ones that we use in computation. In particular, transcendental numbers, those that cannot be finitely constructed, are postulated to fill out the continuum, a concept that itself begs the question of what the transcendental numbers are. The ones that we do encounter, π and e for example, are *names for specific structures*. By tickling the dragon of the continuum, we are abandoning all usual conceptualizations of number, in favor of something else: rare, discrete examples of particular structures that Western math has become interested in amid an uncountable infinite sea of unknowables.

There is another deep problem for Dedekind's definition of number. There is no mechanism for identifying what the natural numbers 1, 2, 3... are. The price that we have paid for our current definition of real numbers is that we have lost the ability to identify the natural numbers! Mathematician Verena Huber-Dyson:

> ...the natural numbers are not "elementarily definable" among the reals; there is no wff of the language of $\mathbb{R}$ [[no well-formed formula, or well-defined

algorithm, in the language of real numbers]] that picks out the natural numbers among the reals.[19]

Calculation

In Figure 11-4 the common operations of arithmetic are expressed as applications of the James axioms, providing a uniform computational model for half-a-dozen apparently different operations. The cost is *many small but similar transformations*. The gain is to be able to see that the diversity in computation rests only on Inversion, Arrangement, and Reflection. Figure 11-4 on the following page uses the explicit unit form of natural numbers (rather than their digit abbreviations) in order to illustrate the fine-grain details of each computation.

Common numeric computation transforms expressions that include operator signs into pure numbers that incorporate no operator signs. This type of computation is possible only for integers being added, subtracted, or multiplied. Unit-fractions such as 1/3 inherently include the divide operation. Many operations, such as square root, $\sqrt{}$, generate real numbers that have infinitely diverse decimal expansions. Every boundary form is also an operation, so the distinction between expressions as objects and expressions as instructions to simplify becomes blurred.

11.3 Remarks

The evolution of types of common numbers has been a series of design choices. The numbers that our culture teaches are not inevitable. In some historical twist we may have ended up with a very different idea of what numbers are. There are viable cultures, for example, that count *one, two, many...*[20] Most cultures do not need or even conceive of real numbers, those creatures that come as an uncountable infinity, each one infinitely complex. We do not emphasize cyclic numbers, even though our clocks show us examples every day.

expression ☞	James form and its reduction	
4 + 2 ☞	oooo oo	
6 ☜	oooooo	put
4 − 2 ☞	oooo <oo>	
2 ☜	oo	cancel
4 × 2 ☞	([oooo][oo])	
	([oooo][o]) ([oooo][o])	disperse
8 ☜	oooo oooo	unmark
4 ÷ 2 ☞	([oooo]<[oo]>)	
	([oo]<[oo]>) ([oo]<[oo]>)	disperse
2 ☜	( ) ( )	cancel
2 ÷ 4 ☞	([oo]<[oo oo]>)	
	([oo]<[([oo][o])([oo][o])]>)	mark
	([oo]<[([oo] [oo])]>)	collect
	([oo]< [oo] [oo] >)	clarify
	([oo]< [oo]> <[oo] >)	split
1/2 ☜	(<[oo] >)	cancel
4^2 ☞	(([[oooo]][oo]))	
	(([[oooo]][o]) ([[oooo]][o]))	disperse
	([oooo] [oooo])	unmark
	([oooo][o])([oooo][o])([oooo][o])([oooo][o])	disperse
16 ☜	oooo oooo oooo oooo	unmark
$\sqrt{4} = 4^{1/2}$ ☞	(([[oo oo]]<[oo]>))	
	(([[([oo][o])([oo][o])]]<[oo]>))	mark
	(([[([oo] [oo])]]<[oo]>))	collect
	(([[oo] [oo]]<[oo]>))	clarify
	(([([[oo]][oo])]<[oo]>))	collect
	(([[oo]][oo] <[oo]>))	clarify
	(([[oo]]))	cancel
2 ☜	oo	clarify

Figure 11-4: *Examples of James arithmetic calculation*

Numbers

Our culture has embraced symbolic notation at the cost of common comprehension. There's nothing wrong with this *per se*. Literature, for example, enriches us in an exclusively symbolic notation. However, according to mathematician Joseph Mazur:

> By the eighteenth century, the language of mathematics was far too symbolized for people to read without a great deal of preliminary tutoring.... the novice had to learn a new visual language while trying to comprehend new material. Understanding such a language either took a very special expertise or enormously intense work persistence. The language was visual, but the meanings were concealed.[21]

A most appealing direction that we could have taken would have been to develop numbers that *make cognitive sense* without particular learning/indoctrination. These numbers may arrange themselves in our minds in spirals. They may come in colors. Larger numbers might be much more inexact: there is 1,000 but no 1,002. And the distance between each pair of integers might decrease with their magnitude. Neurophysiologist Stanislas Dehaene:

> Subjectively speaking, the distance between 8 and 9 is not identical to that between 1 and 2. The "mental ruler" with which we measure numbers is not graduated with regularly spaced marks. It tends to compress larger numbers into smaller space. Our brain represents quantities in a fashion not unlike the logarithmic scale on a slide rule, where equal space is allocated to the interval between 1 and 2, between 2 and 4, or between 4 and 8.[22]

human number line

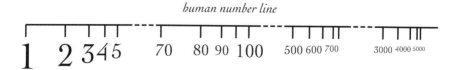

human number line

Chapter 11

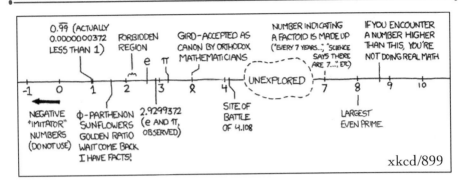

We could live in a culture in which natural numbers were based on cognitive and experiential sensibility rather than on the successor operation and the Inductive Principle. We could elect to design an elementary math curriculum that respected first the number sense of students. Mathematics *could* have been a branch of Psychology.

Rotman, in *Ad Infinitum*, considers the construction of **non-Euclidean arithmetic**, with numbers that get less specific as they grow larger and with addition that incorporates the natural inexactness of large sums. When we say one million, the chances that we mean exactly one million are about one in a million. Therefore

$$1{,}000{,}000 + 1 = 1{,}000{,}000$$

In such a system, we could limit magnitude to always be within the realm of physical possibility, for example

$$\log\log N < 10 \text{ for all } N.$$

$\log\log N = 10$

$\log N = 10^{10}$

$N = 10^{10^{10}}$

$N = 10^{10000000000}$

Capping magnitude, of course, redefines not only the Inductive Principle, but also infinite series, infinite sets, and real numbers. That we treat even the natural numbers as an infinite collection is a *design principle* rather than an obvious fact.[23]

Neuroscientist Dehaene believes that there is a mismatch between our cognitive number sense and the exotic variants of numbers that only recently populate math classes.

Numbers

> What modern mathematicians call "numbers" includes zero, negative integers, fractions, irrational numbers such as π, and complex numbers such as $i = \sqrt{-1}$. Yet, all of these entities, except perhaps the simplest fractions such as ½ and ¼, posed extraordinary conceptual difficulties to mathematicians in centuries past — and they still impose great hardship on today's pupils.... these mathematical entities are so difficult for us to accept, and so defy intuition, because they do not correspond to any preexisting category in our brain.... To function in an intuitive mode, our brain needs images — and as far as number theory is concerned, evolution has endowed us with an intuitive picture only of positive integers.[24]

The common characteristic of each type of number that Dehaene identifies as posing "extraordinarily conceptual difficulties" is that the James form includes an angle-bracket. We go here next.

Chapter 11

Endnotes

1. **opening quote:** H. Weyl (1959) Mathematics and the Laws of Nature. *The Armchair Science Reader.*

2. **one of the most fertile inventions of humanity:** O. Neugebauer (1962) *The Exact Sciences in Antiquity* p.5.

3. **there are three operations, the three boundary types:** James forms appear to be too complicated for common use. Yes, there is an overhead to the boundary notation. Keep in mind that our current notation for numbers is succinct only by virtue of hiding its complexity behind twelve years of elementary and secondary training. The appropriate comparison is not ease of reading but rather ease of use (operations and proof). We will not be dwelling on the many different ways that our culture records numbers. The emphasis is more on how numbers work than on how they are encoded.

4. **the base is placed inside the letter/number for 5:** R. Calinger (1999) *A Contextual History of Mathematics* p.80.

Woody Allen, in H. Eves (1988) *Return to Mathematical Circles*, makes this comment about the numeral 5.

> Standard mathematics has recently been rendered obsolete by the discovery that for years we have been writing the numeral five backward. This has led to reevaluation of counting as a method of getting from one to ten.

5. **a coefficient (a digit in a particular position):** I'll use b for the base rather than the more conventional polynomial variable x, freeing "x" for the multiplication operation.

6. **and possibilities given by conceptual metaphor:** G. Lakoff & R. Núñez (2000) *Where Mathematics Comes From: How the embodied mind brings mathematics into being* p.86. Folks tend to see optimality in structures for which they can envision no alternative.

7. **creates a structure that is parallel rather than sequential:** From the perspective of silicon circuitry, the computational delay in both adders and multipliers is in waiting for the result of a prior process, what kids call "carrying". Sequential adders must wait for each prior digit, their timing

is linear in the number of *possible* carries. Parallel adders exchange space (silicon resources) for time. A conventional parallel adder still must carry, but it does not need to wait when there is no carry coming in. The maximal time to add in such an architecture is the number of *actual* carries.

8. **notation is not neutral to the ideas it expresses:** Stated from the perspective of the Participation Principle, our choice of linear notation reinforces the formation of a linear conceptualization of reality. To put it bluntly, the magnificent invention of place-value comes with a bias against thinking in terms of multiplication and factored forms. A representation that requires adding pieces of a number together encourages a piecemeal perspective, seeing the world divided into components. Tally systems are the extreme of this viewpoint. A representation that has multiplicative interaction as primary encourages seeing the world as an integrated, interconnected system, with every part touching every other part. The Additive and Multiplicative Principles each come with an epistemology. To add is to fuse together perceived fragments that themselves are separate. To multiply is to connect together perceived factors that themselves contain parts of the whole.

9. **before anybody worried about the logic behind them:** I. Stewart (1995) *Nature's Numbers* p.34.

10. **simply beyond the power of algebra to describe:** P. Lockhart (2012) *Measurement* p.68.

11. **no presumption that all real numbers (the uncountable infinity of them) have James forms:** Conventional notation too cannot represent real numbers. We default to naming the *process* by which a theoretical number can be approximated. $\sqrt{2}$ is both irrational and not expressible without the $\sqrt{}$ operation. The real number $\sqrt{2}$ accurately identifies the diagonal of a unit square, but it does not tell us the length of that diagonal any more than the title of a book tells us the story inside. We can convert the abstraction $\sqrt{2}$ into something useful by approximating its value via a computation or a measurement. Within the domain of symbolic mathematics we can manipulate $\sqrt{2}$ as an object whenever it is combined with other symbolic forms that fit established pattern transformation rules. For example, $\sqrt{2} \times \sqrt{2} = 2$. However, note that $2 \times \sqrt{2}$ cannot be further reduced. This implies that natural numbers and real numbers are *not* the same type of number. Contrary to established doctrine, 2 is not a real number just as much as $\sqrt{2}$ is not a whole number.

More commonly, mathematicians identify an infinite sequence which begins with a pattern and ends with an ellipsis. Although completely legitimate for mathematics, an appeal to abstract infinities does not particularly benefit students who want to learn how to *use* mathematics. It's an open question whether or not we can tell the difference between algebraic and transcendental reals by looking at their James structure.

12. **define irrational numbers by means of the rational numbers alone:** R. Dedekind (1872) *Continuity and Irrational Numbers* p.10.

13. **passing from an already-formed individual to the consecutive new one to be formed:** Dedekind, p.4.

14. **Even these words are used to construct an algorithm:** L. Wittgenstein (1933) *Philosophical Grammar* §468.

15. **no reason or criterion "for irrational numbers being complete":** L. Wittgenstein (1930) *Philosophical Remarks* §181.

16. **but rather, as conventionally adopted useful fictions:** M. Leng (2010) *Mathematics and Reality* p.261. Leng continues her analysis of mathematical fictionalism:

> ...there is nothing in our current best scientific worldview that gives us reason to believe in the distinctive mathematical objects (such as numbers, functions, and sets) posited by many of our ordinary mathematical theories.

17. **controlling what mathematicians can meaningfully and usefully write down:** B. Rotman (2000) *Mathematics as Sign: writing imagining counting* p.122.

18. **Existence, what we call reality, is built on the discrete:** J. Wheeler, (1988) World as system self-synthesized by quantum networking. *IBM Journal of Research and Development* **32**(1) p.6. Also S. Wolfram (2002) *A New Kind of Science*.

19. **that picks out the natural numbers among the reals:** V. Huber-Dyson (1998) On the Nature of Mathematical Concepts: Why and How Do Mathematicians Jump to Conclusions? in *Edge* p.308.

20. **viable cultures, for example, that count one, two, many:** For example, the Australian aboriginal Warlpiris and the Amazonian Pirahã. This does not imply that these and other cultures do not perceive, for example, exactly three, nor does it imply that there are not other words in the language that provide for others types of quantification. The contrast is that a successor function that builds numbers one-by-one (so to speak) may not be necessary or even useful.

21. **The language was visual, but the meanings were concealed:** J. Mazur (2014) *Enlightening Symbols* p.179-80.

22. **between 2 and 4, or between 4 and 8:** S. Dehaene (2011) *The Number Sense: How the mind creates mathematics* p.65.

23. **a design principle rather than an obvious fact:** In another formal approach to conforming whole numbers to human use, Sazonov limits the largest number to 2^{1000}.

V. Sazonov (1995) On Feasible Numbers in D. Leivant (ed.) *Logic and Computational Complexity* LNCS 960 p.30-51. Online 12/17 at http://www.csc.liv.ac.uk/~sazonov

The point of humanized numbers is made by the following anecdote. A school child asked the museum curator how old the dinosaur fossil was. The curator said sixty-five million years old. When the child returned to the museum two years later, she asked to curator if she could see the dinosaur fossil that was sixty-five million and two years old.

24. **evolution has endowed us with an intuitive picture only of positive integers:** Dehaene, p.75-76.

number line comic: xkcd cartoon by Randall Munroe. Permanent link online 2/17 at http://xkcd.com/899/

Chapter 11

Chapter 12

Extension

To treat numbers as unique, eternal, and unchangeable is a kind of number mysticism reminiscent of the legendary Pythagoreans.[1]
— *Alberto Martínez (2012)*

Now we'll extend the James patterns into one of the more challenging areas of 4th grade arithmetic — fractions. We'll first examine the form of Accumulation to identify types of cardinality. Then we'll generalize Arrangement to include fractions and ratios. Later we'll also extend the concept of exponential and logarithmic bases and the Reflection axiom.

12.1 Cardinality

The contents of a container can be assigned a cardinal number N that corresponds to the number of forms within that container. Cardinality can be thought of as a tally that is associated with a natural number. The Replication rule gives specific permission to substitute a cardinality N for an ensemble of tally marks. Cardinality can be generalized to other types of ensembles. For ease of reading, I'll use the multiplicative version of Replication.

replication
$([A][N]) = A..._N..A$

Chapter 12

Here is multiplication converted to additive replication.

$$A \times N \quad ☞ \quad ([A][N])$$

substitute $\quad\quad ([A][o..._N..o])$

replicate $\quad\quad A..._N..A \quad\quad ☞ \quad A +..._N..+ A$

We've also seen replication of [A] forms. A collection of square bracket forms can be interpreted as parallel multiplication as well as parallel addition of logarithmic forms.

$$A^N \quad ☞ \quad (([[A]][N]))$$

substitute $\quad\quad (([[A]][o..._N..o]))$

replicate $\quad\quad ([A]..._N..[A]) \quad ☞ \quad A \times..._N..\times A$

Naturally, any form can be replicated. Replicas generated by Replication can be counted by natural numbers. By abandoning the 1, 2, 3,... of natural number counting, we can generalize the forms of multiplication and exponentiation to accommodate any type of numeric count, including negative, fractional and irrational counts. From the James perspective, N can stand in place of any numeric form. Dispersal into units, however, is no longer available for non-rational counts. This is a significant loss since Arrangement is the only axiom that permits rearrangement of forms.

Should we count with negative numbers, the form would be ([A][<o>]), which can be interpreted as a positive number times negative one. Colloquially, a negative count might be thought of as counting backwards. Here is a demonstration that a negative number and counting by negative units have equivalent structures.

$$-A \quad ☞ \quad < A \quad >$$

mark $\quad\quad <([A][o])>$

demote $\quad\quad ([A][<o>]) \quad ☞ \quad -1 \times A$

An application of Promotion shows that the polarity of multiplication can be converted from a global to a local property. The notorious concept of a negative times a negative disappears via a change in perspective, from contents to context. In general Promotion can lift angle-brackets to a shallower, more pervasive context.[2]

([<A>][<B>])
<<([A][B])>>
([A][B])

Another use of Promotion is the conversion of negative powers to unit fractions.

```
(   ([[A]][   <N>   ])   )   ☞   A⁻ᴺ
(<  ([[A]][    N    ])  >)   ☞   #⁻⁽log_#A × N⁾         promote
(<[((([[A]][   N   ]))]>)    ☞   1/Aᴺ                   enfold
```

The concept of cardinality is generalized by extending the ability to tally using any type of numbers. Not only can we count three twos and three πs, we can "count" two threes and π threes. It may make sense to accumulate three is, and it may make sense to employ 3 as a way to increase the magnitude of i threefold, but does it make sense to accumulate an imaginary number of threes? What exactly does "i threes" mean?

Fortunately, within the James form, we do not need to interpret the structure of cardinality as multiplication or as counting. You might say that multiplication and counting are both *applications* of a more general boundary structure. We can explore the structural consequences of boundary arithmetic without being concerned that they make sense within conventional arithmetic. Boundary transformations do, of course, maintain consistency. What is being generalized is not the idea of counting nor the concept of cardinality. We are expanding our conceptualization of James patterns, particularly James frames.

The Replication theorem supports a tally of forms as well as a tally of units. In the case of counting-by-twos introduced in Chapter 9, we associated two units with each tally. In effect each tally counted as two. We can also associate an arbitrary value, such as 15.37 or π or √2, with each tally mark.

Chapter 12

count by	expression	written as	form
zero	0 x A	0	([ ] [A])
units	1 x A	A	([o] [A])
reciprocals	1 / A	A⁻¹	([o]<[A]>)
negatives	−1 x A	−A	([<o>] [A])
irrationals	π x A	πA	([π] [A])
imaginaries	i x A	Ai	([i] [A])

Figure 12-1: *Types of cardinality*

And to be complete, zero cardinality is one interpretation of the Dominion theorem, which we could equally interpret as multiplying by zero, as counting by zero, and as the non-existence of a particular form.

$$0 \quad ☞ \quad void$$

emit $\quad$ ([A][ ]) ☞ 0 x A

[A] is absorbed by Dominion making it indeterminate. Deleting [A] yields Void Inversion.

$$(\ [\ ]) \quad ☞ \quad 0$$

Figure 12-1 shows several interpretations of the form of cardinality. For convenience, the figure shows hybrid boundary forms. We can construct both Indication instances and Replication instances of units and negative units, of rationals and irrationals, and of imaginaries.

Any contents of a round-bracket can be cast into the form of a frame, since

enfold $\quad$ (A B) = (A [(B)])

That is, the contents of the round boundary can be read as participating in multiplication *or* in addition. We thus have the choice of abandoning or extending the concept of cardinality to include arbitrary forms.

Extension

Depth-value

As defined in Chapter 11 the double-struck depth-value boundary is an *abbreviation* for a particular frame, the **unit magnitude frame**.

$$((A)) = ([A] \ [base])$$

We'll remove the invasive word "base" by substituting its boundary form.

$$\begin{array}{ll}([A] \ [\ base\ ]) & \\ ([A] \ [((\ ))]) & \text{substitute} \\ ([A] \quad o \quad) =_{def}= ((A)) & \text{clarify}\end{array}$$

Here's an example,

$$\begin{array}{ll}432_{base\text{-}10} \ \mathrel{\mathop{\Rightarrow}} \ ((\ (\ 4\ )\ 3\ )\ 2 & \\ ((\ ([4]\ o)\ 3\ )\ 2 & \text{substitute} \\ ([([4]\ o)\ 3]\ o)\ 2 & \text{substitute}\end{array}$$

The unit-magnitude frame is a special case of the generic magnitude frame discussed in Chapters 8 and 11.

$$((A)) = ([N]\ o) \quad \mathrel{\mathop{\Rightarrow}} \quad N \times base \qquad \text{\textbf{unit magnitude frame}}$$

12.2 Fractions

Fractions have been around since the ancient Egyptians, but four thousand years ago the concept of fraction was a **unit fraction**, the reciprocal of a natural number, $1/n$. Transformation of fractions today is notorious, the most common error found in elementary math classes.

A conventional fraction such as $2/3$ multiplies a number (here 2) by a unit fraction (here $1/3$). In James arithmetic, multiplication and division share the same structure but for an angle-bracket that identifies divisors.

$$\begin{array}{l} A \times B \ \mathrel{\mathop{\Rightarrow}} \ ([A]\ [B]\ ) \\ A \div B \ \mathrel{\mathop{\Rightarrow}} \ ([A]{<}[B]{>}) \end{array}$$

Chapter 12

When A = 0, we get the generic form of a unit fraction, (<[B]>).

$$1/B \; ☞ \; ([o] <[B]>)$$
clarify
$$(\quad\;\; <[B]>)$$

Multiplying A by 1/B clarifies to the form of division.

$$A \times 1/B \; ☞ \; ([A][(<[B]>)])$$
clarify
$$([A] \;\; <[B]> \;)$$

The reciprocal of a reciprocal acts appropriately

$$1 \div 1/B \; ☞ \; ([o]<[(<[B]>)]>)$$
clarify
$$(\quad < \;\; <[B]> \;\; >)$$
unwrap
$$(\qquad\;\; [B] \qquad)$$
clarify
$$B$$

Conventionally, unit fractions can also be expressed in exponential notation.

$$B^{-1} \; ☞ \; (\;([[B]][<o>]) \;)$$
promote
$$(<([[B]][\;o\;])>)$$
unmark
$$(< \quad [B] \qquad\quad >) \quad ☞ \; 1/B$$

Parallel Multiplication

The general form of multiplication accommodates any number of forms being multiplied together.

parallel multiplication
$$A \times ..{}_N.. \times Z \; ☞ \; ([A]..{}_N..[Z])$$

Thus, a James fraction does not necessarily need to be reduced. It can include any number of numerators and any number of denominators.

$$(A \times B)/(C \times D) \; ☞ \; ([A][B]<[C]><[D]>)$$

Alternatively, we could combine the two denominator forms above within a single angle-bracket:

$$([A][B]<[C]><[D]>)$$
join
$$([A][B]<[C]\;\;[D]>)$$

Extension

In general, a James fraction takes this form

$$([A]..._M..[K] <[L]..._N..[Z]>)$$ **parallel fraction**

The multiplicative rules for common fractions are incorporated into the form of the fractions themselves. The form of parallel multiplication supports a factored notation for numbers.[3] The choice to permit multiple numerators and denominators comes from the availability of parallel processing. Multiple square-bracketed forms $[A]..._M..[K]$ share the Additive Principle, their joint sum is represented by the ensemble of the components.

Reducing Fractions

Reducing or constructing the fraction A/A is quite natural. Note how the potential of A/A is embodied within the structure of one, ( ).

```
      A/A  ☞  ([A]<[A]>)
             (           )                    cancel

     2A/A  ☞  ([2][A]<[A]>)
             ([2]         )                   cancel
                2                             clarify

     6A/2  ☞  ([   6   ][A]<[2]>)
             ([([3][2])][A]<[2]>)             substitute
             (  [3][2]   [A]<[2]>)            clarify
             (  [3]      [A]    ) ☞ 3 x A     cancel
```

Adding Fractions

As you might expect, addition of fractions relies heavily upon Arrangement. Arrangement can be generalized to accommodate a wider diversity of patterns, particularly those patterns that occur as special cases of adding fractions. Figure 12-2 shows various theorems that adapt Arrangement to different types of conventional fractions that we use. Each new form of Arrangement is a theorem,

Chapter 12

number type	expression ☞	form of arrangement
unit fraction	1/B + 1/D	([B D]<[B][D]>)
compound fraction	A + C/D	([C ([A][D])] <[D]>)
fraction	A/B + C/D	([([A][D])([B][C])]<[B][D]>)

Figure 12-2: *Customized versions of arrangement as addition of fractions*

a few transformation steps away from the Arrangement axiom. Each identifies a particular type of fraction addition. Here are three examples of applying Arrangement to manipulate factions.

Add one to a fraction

```
                A/B + 1  ☞  ([A]<[B]>) (               )
     create                 ([A]<[B]>) ([B]<[B]>)
     collect                ([A B]<[B]>)
                                        ☞  (A + B)/B
```

Add two fractions with a common denominator

```
                A/C + B/C  ☞  ([A]<[C]>) ([B]<[C]>)
     collect                  ([A B]<[C]>)  ☞  (A + B)/C
```

Add two fractions without a common denominator

```
                A/B + C/D  ☞
              (    [A]           <[B]>) (    [C]              <[D]>)
   create     (    [A][D]  <[D]><[B]>) (    [C][B]     <[B]><[D]>)
   join       (    [A][D]  <[D]  [B]>) (    [C][B]     <[B]  [D]>)
   enfold     ([([A][D])]<[D]   [B]>) ([([C][B])]<[B]   [D]>)
   collect    ([([A][D])([C][B])] <[D][B]>)
                                        ☞  ((A x D) + (C x B))/(D x B)
```

In the last example there is nothing common to collect. This then suggests a strategy of *constructing communality* by inserting reflection pairs. Reflection is particularly easy to use, we bring into existence precisely the communal forms that are needed. The communal forms themselves are defined by a desire to collect.

These steps are associated with the way that manipulation of fractions is represented and taught, not necessarily the way that fractions are used. But there is a deep ambiguity. We return to the question asked in Chapter 6. Is 1/3 one number or two? It is both depending upon context. The fraction 1/2, as in one-half of the cake, is different that the ratio 1/2, as in 1 out of 2 students understand fractions.

Formally a **fraction** is an ordered pair, it is two numbers. The fraction 3/4 identifies a creature for which adding 4 of them together, yields 3. A **rational number** is an infinite equivalence set consisting of all ordered pairs $\langle a,b \rangle$ that reduce to the same value.

fraction as ordered pair

$a/b = \langle a,b \rangle$

Rational numbers come with rules, the primary one being that the second of an ordered pair (the denominator) cannot be zero. Of course, this is the *standard* definition. In Volume III we will encounter and use forms that can be interpreted as a violation of the divide-by-zero restriction.

Compound Ratios

The meaning of a compound fraction is very difficult to explain to students. **Compound fractions** are ratios of ratios, but they are often vilified as the symbolic division of a fraction by a fraction.

(A/B) ÷ (C/D) = (A/B) x (D/C) = (A x D)/(B x C)

In boundary form,

```
(A/B) ÷ (C/D)  ☞  ([([A]<[B]>)]<[([C]<[D]>)]>)
                  (  [A]<[B]>    <   [C]<[D]>  >)    clarify
                  (  [A]<[B]>    <   [C]>[D]   )     react
                  (  [A]<[B]         [C]>[D]   )     join
```

Chapter 12

Why is dividing two fractions the same as multiplying by the reciprocal of the denominator? We can see from the transformations on angle-brackets that <[D]> is contained by a second angle-bracket. The Reaction theorem deletes the two angle-brackets enclosing [D], bringing it into what we call the numerator. At the same time, Reaction relocates [C] to the denominator.

Adding Ratios

Leonardo of Pisa (aka Fibonacci) incorporated the fraction line to distinguish numerator from denominator in his 1202 *Book of Calculation*.[4] These fractions represented division into parts. It was Leibniz (1693) who advocated that the sign of ratio and proportion should be the same as the sign of division.[5] Napier's logarithmic tables in the seventeenth century popularized decimals as a way to achieve multiplication via addition. These decimals were not ratios. As the use of decimals gained in popularity in the late nineteenth century, *the rules of fraction addition were changed*, from addition of ratios to addition of decimals that might be interpreted as fractions. Fractions originated as ratios, yet, in the words of Keith Devlin,

> ...the mathematical community long ago decided that adding fractions means something different from adding proportions — even though fractions may be and are used to quantify proportions.[6]

If 1/2 of my two cats is white, and 1/3 of your three cats is white, then 2/5 of our five cats is white. Here fractions are used cumulatively, as proportions. They are neither added nor multiplied by conventional rules of fractions. If we were to say that .5 of my cats is white, while .33 of your cats is white, we can no longer find the proportion of white cats. An average is not a proportion.

$$(.5 + .33)/2 = .42 \neq .40 = 2/5$$

Extension

How would we generalize Arrangement to accommodate ratios? The addition rule for *ratios* is

A/C + B/D ☞ ([A]<[C]>) ([B]<[D]>)
                     ([A B]<[C D]>)   collect ratios
                             ☞ (A + B)/(C + D)

This type of addition resembles addition of matrices, and would require a redefinition of the Arrangement axiom.

12.3 Bases

Is the use of the arbitrary base # valid in all circumstances? From the perspective of James algebra, YES because there is no concept of a base. Bases are in the interpretation. Does the idea of an arbitrary base conform to conventional usage? Does James algebra make any particular interpretations inaccessible? Conventionally, we should be safe with any positive real number as a base for an exponent, and we should be safe raising that number to any numeric power. However, if powers are not integers (for example $2^{1.84}$ or 2^π), then the result is conventionally interpreted as a complex number. We address the exponents and logarithms of complex numbers in Volume III.

$(1/6)^N$
$(.345)^N$
$(\sqrt{2})^N$
$(\pi)^N$

Logarithmic Base

Figure 12-3 shows the variety of forms for logarithms with different bases. There are three cases marked with an arrow for which the form of the logarithm is as yet undefined: ∞, 1, and 0. Each incorporates an empty square-bracket. These exceptions are also addressed in Volume III.

Let's start off with conversion between different logarithmic bases. This formula arises naturally from the base-free perspective of James form.

Chapter 12

base	expression	form	
N = ∞	log$_∞$ A	(<[[<[ ]>]]>[[A]])	←
1 < N < ∞	log$_N$ A	(<[[N]]>[[A]])	
N = 1	log$_1$ A	(<[ ]>[[A]])	←
0 < N < 1	log$_{1/N}$ A	(<[[(<[N]>)]]>[[A]])	
	− log$_N$ A	<(<[[N]]>[[A]])>	
N = 0	log$_0$ A	(<[[ ]]>[[A]])	←
N < 0	log$_{-N}$ A	(<[[<N>]]>[[A]])	

Figure 12-3: *Logarithms with a range of bases*

Logarithm conversion formula

$$\log_C A = \log_C B \times \log_B A$$

$\log_C B \times \log_B A$ ☞

```
              ([ (<[[C]]>[[B]]) ][ (<[[B]]>[[A]]) ])
clarify       (    <[[C]]>[[B]]      <[[B]]>[[A]]    )
cancel        (    <[[C]]>                   [[A]]   )
```
☞ $\log_C A$

The logarithm to a *specific base* (B above) can be stated as a ratio between two logarithms, each having a generic base #. The constant of proportionality, 1/log$_\#$ B, converts one base to any other base.

$$\log_B A = \log_\# A \times 1/\log_\# B$$

$\log_B A$ ☞   ([[A]]   <[[B]]>   )
enfold   ([[A]] [(<[[B]]>)])
hybrid   [A] × (<[[B]]>)
hybrid   [A] ×   1/[B]
☞ $\log_\# A \times 1/\log_\# B$

This means that specific bases are not essential for transforming logarithmic forms.[7] Transformation between bases consists of clarifying (via Inversion) and reducing (via Reflection) the original base.

Extension

Of the two different James derivations of the logarithm conversion formula, the first shows how $\log_B$ is eliminated from a result via Reflection. The second shows how $\log_B$ is converted into an arbitrary base via clarification. Both are too elaborate. The form of a logarithm can be *read* as the form of division, providing direct access to the base conversion formula.

$$\log_B A \ \mathrel{\mathop{\mathrm{☞}}} \ (\ [[A]] \ <[[B]]> \ )$$
$$[A] \ \div \ [B] \qquad \text{hybrid}$$
$$\mathrel{\mathop{\mathrm{☞}}} \ \log_\# A \div \log_\# B$$

Logs and Powers

By definition,

$$\log_B A = R \quad \Leftrightarrow \quad B^R = A$$

Converting each equation into boundary form, we have

$$\log_B A = R \ \mathrel{\mathop{\mathrm{☞}}} \ (<[[B]]>[[A]]) \ = R$$
$$B^R = A \ \mathrel{\mathop{\mathrm{☞}}} \ ((\ [[B]] \ \ [R]\ )) = A$$

Here, the base of the logarithm is explicit. When B itself takes the form (( )), base-B is absorbed into the structure of the round bracket. Here's the structural hierarchy.

$$(\ ) \ \mathrel{\mathop{\mathrm{☞}}} \ B^0 = 1$$
$$((\ )) \ \mathrel{\mathop{\mathrm{☞}}} \ B^1 = B$$
$$(\ R\ ) \ \mathrel{\mathop{\mathrm{☞}}} \ B^R$$

Therefore

$$\log_B A \ \mathrel{\mathop{\mathrm{☞}}} \ (<[[\ B\ ]]>[[A]])$$
$$(<[[((\ ))]]>[[A]]) \qquad \text{substitute}$$
$$[A] \qquad \text{clarify}$$

This shows that the interpretation of ( ) as a power operation and [ ] as a logarithm operation aligns. So long as the specific base is assigned to the round-bracket, then all boundary forms will conform to that same base. If *no base* is absorbed into ( ), then the mechanism of James forms remains the same, without concern for interpretation in a base system or interpretation as powers and logarithms.

Fractional Bases

We'll consider a generic fractional base, M/N. When M > N we have a generic rational base greater than one, and when M < N, the generic rational base less than one.

$$\log_B A \;\;\text{☞}$$

	(<[[B]]>[[A]])
hybrid	(<[[M/N]]>[[A]])
substitute	(<[[([M]<[N]>)]]>[[A]]) ☞ $\log_{M/N} A$
clarify	(<[[M]<[N]>]>[[A]])

$$\text{☞}\;\; \log_\# A / (\log_\# M - \log_\# N)$$

For the last interpretation above, we have taken advantage of the arbitrary base # to *read* the form as a division ([X]<[Y]>) with X = [A] and Y = [M]<[N]>.

The result is a permissible but unconventional form with special cases. When N = 1, we get a conventional whole number base. When M = 1, we get an unconventional unit fraction base. Here's how a unit fraction base works.

hybrid	(<[[M/N]]>[[A]])
substitute M=1	(<[[([o]<[N]>)]]>[[A]])
unmark	(<[<[N]>]>[[A]]) ☞ $\log_{1/N} A$
promote	<(<[[N]]>[[A]]> ☞ $-\log_N A$
demote	(<[[N]]>[<[A]>])
enfold	(<[[N]]>[[(<[A]>)]]) ☞ $\log_N 1/A$

An interesting result. A unit fraction base-1/N logarithm is the same as the logarithm of the reciprocal in base-N. Conventionally

$$\log_{1/N} A = -\log_N A = \log_N (1/A)$$

Notice that the logarithm of a unit fraction is negative. We'll revisit this equation visually in Chapter 13.

Base-e

The natural logarithmic base, e, is a transcendental number, it cannot be accessed via polynomial equations.

Extension

e does show up, however, throughout calculus, and is generally defined as the integral of 1/x. e can also be defined by a limit process. At this point we have introduced no mechanisms for addressing the theory of limits.

$e = \lim_{n \to 0}(1+n)^{1/n}$ ☞
                (([[o n]]<[n]>)) as n→0

Alternatively, the limit definition of e can be written as

$e = \lim_{n \to \infty}(1+1/n)^n$ ☞
                (([[o (<[n]>)]][n])) as n→∞

Without concern for its derivation, we can use e as an embedded base, # = e. This places a direct representation of e into James forms.

      ( )   ☞ $e^0 = 1$
      (( )) ☞ $e^{e^0} = e^1 = e$
      ((( ))) ☞ $e^{e^{e^0}} = e^e$

12.4 Fractional Reflection

Until now, we have been treating boundaries themselves as unitary. Given the round boundary as a unit, one-half of a round boundary is expressed as a unit fraction.

        1/2 ☞ ([o]<[2]>)
                (<[2]>)   clarify

The modification of the form of the round boundary takes place within the interior of the boundary. This works because the round-bracket is numeric. It would be inappropriate, of course, to formulate a fraction of the square boundary since it is non-numeric.

But what is one-half of an angle boundary? What is a fraction of a reflection? Should we translate this question into conventional terms, we would be asking about a fraction of negation, or a fraction of division.

Chapter 12

The angle boundary has an alternative form. In Chapter 10 the J-transparency theorem provided an example of boundary transparency.

J-transparency $\quad$ [<(A)>] = [<( )>] A

Here's the demonstration of J-transparency.

	[<(A)>]
enfold	[([<(A)>][o])]
promote	[<[(A)][o])>]
demote	[([(A)][<o>])]
clarify	A [<o>]

J-transparency allows us to convert all angle brackets into a frame with J as the frame-type.

$$<A> = (J\ [A])$$

	< A >
enfold	([<([A])>])
J-transparency	([<( )>][A])
substitute	(J [A])

The angle-bracket is thus an *abbreviation* for a particular type of frame, the **J-conversion frame**. By reformulating the angle-bracket <A> as (J [A]), we can convert the James calculus into a *two* boundary system augmented by a single constant J.

The question of half of reflection become tractable as an analogous question, what is half of J? This time we will approach the construction from the conventional notation of the imaginary number i.

$$\sqrt{-1} \ \ \Rightarrow \ \ (([<o>]<[oo]>))$$

substitute	((J <[oo]>))
hybrid	(J/2) $\Rightarrow$ #^(J/2)

To expose the value of J/2, we can take the logarithm.

$$\log_{\#} \sqrt{-1} \ \ \Rightarrow \ \ [(J/2)]$$

clarify	J/2

302

One-half of J is the logarithm of i. In Volume III, we will explore the form of J, with some surprising results. For example, J is a second *numeric* form other than 0 that is equal to its own negation!

$$[<o>] = <[<o>]>$$

I can't resist including the demonstration. The technique is the same as the proof of Dominion in Chapter 10.

```
<       [<o>]>
<       [<o>]><[<o>]    > [<o>]           create
<       [<o>]  [<o>]    > [<o>]           join
<[  ([<o>]    [<o>])  ]> [<o>]            enfold
<[<<([ o ]    [ o ])>>]> [<o>]            promote
<[  ([ o ]    [ o ])  ]> [<o>]            unwrap
<                       > [<o>]           clarify
                          [<o>]           void reflect
```

With J ≠ 0, we have both

| 0 = -0 | *and* | 0 + 0 = 0 |
| J = -J | *and* | J + J = 0 |

12.5 Remarks

We've come to the end of the integration of common and James arithmetic. In the next two chapters, we'll return to James basics to explore some of the sensual and experiential dialects. This diversity of representation is considered to be different *notations* in a string-based vocabulary. The word *dialect* is perhaps too close to diversity in language, while the word *notation* is probably too close to diversity in symbol strings. In the pre-digital age, perceptual and dimensional diversity were associated with types of media, such as radio, television, and live performance. After digital convergence, the substrate of all media is the same, only the output devices differ.

We will next explore a diversity of methods to understand boundary thinking through different sensory modalities, particularly the visual, the tactile, and the experiential.

Chapter 12

Endnotes

1. **opening quote:** A. Martínez (2012) *The Cult of Pythagoras* p.186.

2. **Promotion can lift angle-brackets to a shallower, more pervasive context:** The use of Promotion does not resolve the initial design decision about multiplication of negative numbers discussed in Chapter 2. Promotion is a consequence of a particular decision to accept Reflection as an axiom.

3. **parallel multiplication supports a factored notation for numbers:** Why do we prefer the expression of a fraction in its lowest common denominator? Why is 1/2 preferable to 2/4 or 6/12 or 50/100? More specifically, why is it preferable to exert the computational effort to convert a natural fraction such as half-a-dozen, to its lowest common denominator, one-half dozen? Are the chances 50% or 1/2?

A parallel fraction such as (4/52)x(4/52) tells us the chances of drawing an ace twice in a row out of a deck of 52 cards. These chances reduce to 1/169. If it were relevant to keep track of the size of deck of cards (the **sample space** of the draw), then the reduced fraction robs us of that information. Similarly if two of my four cats are sick, it may be more beneficial to tell the vet that two out of four cats are sick, rather than to say that half of my cats are sick.

4. **the fraction line in his 1202 *Book of Calculation*:** F. Cajori (1928) *A History of Mathematical Notations* Book I §235.

5. **should be the same as the sign of division:** Cajori, §259.

6. **fractions may be and are used to quantify proportions:** K. Devlin (2011) *Mathematics Education for a New Era* p.68.

7. **not essential for transforming logarithmic forms:** Similarly, in an additive rather than a multiplicative context, the distance between all numbers is the same. The scale of that distance is not essential for addition to proceed.

Chapter 13

Dialects

Is it unnatural or deviant to suggest that immersion in a virtually realized mathematical structure...be the basis for mathematical proofs?[1]
—Brian Rotman (2000)

One great advantage of boundary languages is that one-, two-, and three-dimensional forms are available as equivalent dialects. This chapter is primarily a sensory exploration of postsymbolic forms of James algebra.[2] Some postsymbolic representations for depth-value were considered in Chapter 4. The main idea is that the foundations of arithmetic and algebra are not necessarily abstract, rather they can be anchored to our physical experience *not by interpretation but by structure*. The **Principle of Participation** dominates this chapter: the representation of formal concepts interacts strongly with how we think.[3]

An algebra is defined by its transformations, not by the notation that expresses these transformations. However, since the boundary dialects range over symbols, icons, images, manipulatives and experiences, each dialect has nuances that lead to different ways of thinking about both arithmetic and algebra. Each dialect supports a different physical concept, converting the inside/outside contains relation into surrounding, holding, relational, territorial, gravitational, stackable, inhabitable and traversable environments. All dialects hold in common that the containment relations between the visual elements are invariant. Spatial dialects are not analogous to different spoken languages, such as English, Latin or Farsi.[4] Forms are more reminiscent of recalling different but similar experiences. When we think of a house, for example, there are many quite different exemplars that engender the same generic idea of dwelling. Since formal mathematics has been dominated by string representations for over a century, there are no analogous mathematical languages that resemble the expressive variety of spatial dialects. Perhaps the closest is the diversity of visual layout for graphs and networks. This chapter moves symbolic strings into new perceptual and cognitive territory, enhancing their intended meaning while maintaining their formal rigor and, most importantly, while returning simple mathematical concepts to mundane experience.

13.1 Postsymbolic Form

The early twentieth century founders of modern formality in mathematics began rightfully with the most degenerate form, one dimensional strings of symbols. They were after all attempting to put any sound foundation underneath the mess they found themselves with after the loss of certainty in the nineteenth century. Symbolic formalism, however, was conceived in an era before telephones that are cameras, before MTV, before TV, even before radio, before airplanes and automobiles, before public education, before World Wars. Before computers, before digital communication, before electronic calculators, before pervasive electricity, before you and I were born.

Symbolic strings were the obvious tool for expressing the formal structure of mathematics, basically because other tools did not yet exist. A century later we can be brave enough to move to (at least) two dimensions, without fear of confusion and without loss of mathematical perspective. Mathematician John Littlewood:

> A heavy warning used to be given that pictures are not rigorous; this has never had its bluff called and has permanently frightened its victims into playing for safety.... But pictorial *arguments*, while not so purely conventional, can be quite legitimate.[5]

Our agenda has been to lift natural numbers, what many would consider the foundation of mathematics, out of their presumed symbolic confinement. Mathematics is the study of pattern, but if it is to be a human endeavor, those patterns must be broader than strings of characters. Here's Keith Devlin:

> Because these patterns are, for the most part, highly abstract, their description and study require an abstract notation....The complexity and abstraction of most mathematical patterns make anything other than symbolic notation prohibitively cumbersome to use.[6]

What is now needed in our shared exploration are some examples of higher dimensional calculations and demonstrations that are not clumsy or cumbersome. Spatial dialects are not necessarily an explanation of how physical reality works but instead an exploration of how common arithmetic works.

Dimensionality

The fundamental shift into postsymbolic thinking is to engage higher dimensions. To expand from making a point, to drawing a line, to laying out on a surface, to molding into a shape, to experiencing a world.

Here is a listing of the dimensionality of the dialects in this chapter.

- **1D textual linear:** delimiting brackets
- **2D diagrammatic:** closed curves, networks, maps
- **3D environmental:** buckets, blocks, walls
- **4D experiential:** paths, rooms

This list identifies the primary dimension of each representation, not every dimensional perspective that is available. This is not to ignore, for example, that bounding curves can also be converted into bounding surfaces.

One-dimensional forms fracture containment into left and right brackets. Two-dimensional forms maintain planar containment but require replication of objects. They do not fully support **structure sharing**, in which different objects with matching parts merge their parts. Three-dimensional forms can accommodate the entire diversity of types of connectivity. Forms of any spatial dimension can accommodate the addition of time, which leads to animation and dynamics.

From any dimension, all lower dimensions can be reached by degrading some features of the representation. The process of **dimensional reduction** both introduces unintended artifacts that distort interpretation, and removes experiential richness that could potentially enhance comprehension and embodiment. As our standard example, commutativity is a property of linear forms of expression, an awareness of which is certainly necessary for non-commutative spoken words, musical notes and temporal experiences. The group theoretic rule of commutativity gives us permission to consider forms of expression that are not bound in space, in time or in sequence. In contrast diagrams do not need a concept of commutativity, they have instead the space of the page they are recorded upon. A diagram by its very nature gives us permission to consider forms of expression that are not limited to a line or a sequence.

Independent of dimensionality, boundary forms must include a direction or *orientation* to indicate the deepest and the shallowest nested form. Depth-value includes ordinal numbering that corresponds to depth of nesting. In some applications, for example those that incorporate semipermeable boundaries, being able to count how deep is unnecessary. In the James form, the polarity of depth (odd or even) is important. It is, however, the perspective of the reader that defines the fundamental orientation of forms, including the location of the origin, which in typography is declared by the curvature of the pair of brackets.

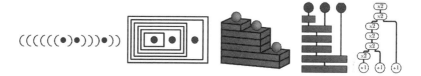

Varieties

There are three major classes of structure that distinguish boundary forms. **String** varieties enlist encoded characters or sequences of characters, while embedding structure in their ordering. **Geometric** varieties enlist different metric structures that share relationships across their measures. Changing a metric is analogous to changing a symbol in a string language. **Relational** varieties enlist different spatial structures that embody different tangles of relations. Geometric varieties are topological forms with rubber-sheet geometries, while relational varieties are not limited by the rubber-sheet. The appearance taken by the relational varieties can be generated by extruding and rotating into a higher dimensional space, by converting links to borders, by exchanging objects for processes and by structure sharing. Some varieties of form highlight the neutral background space to delineate structure. For example, the space between typographical words is used to identify grouping but is not taken as a grouping operator.

Chapter 13

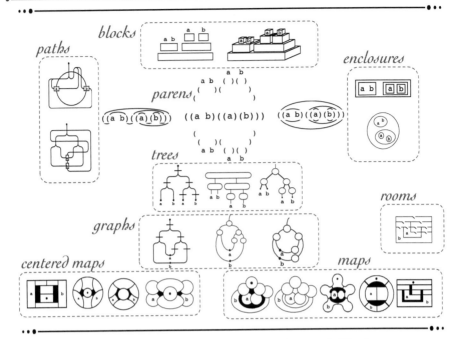

Figure 13-1: *Roadmap for generating spatial dialects*

A dominant characteristic of boundary forms is the **point-of-view** of the reader/participant. Forms can be read from the outside, objectively, or from the inside, subjectively. Subjective reading includes participation within the form itself. Linear form, in contrast, lacks an inside, forcing the perspective that the reader is outside, in some higher dimension. In its desire to remove human bias, mathematics has embraced the outside, objective viewpoint, creating a notation that lacks both participation and dynamics. Process must then be exhibited as *steps*. A refinement that appears to have been overlooked is that objectivity, seeing things as objects, viewing reality from the outside, does not achieve neutrality. Objectivity limits our perspective so severely that we believe we are not only super-human but that we have access to locations outside of our universe!

Objectivity makes us the outermost boundary.

Dialects

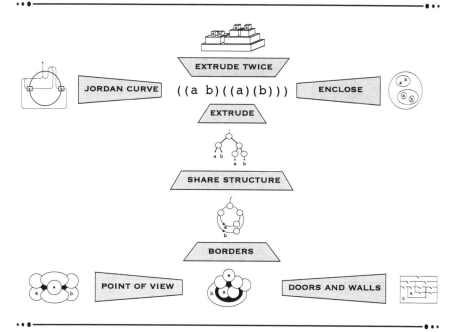

Figure 13-2: *Structural transformations for generating spatial dialects*

13.2 Roadmap

Figure 13-1 provides a roadmap for transcription between the delimiting bracket textual forms and several types of diagrammatic, spatial and experiential representation. The roadmap includes only one type of boundary, it does not address the modeling complexities introduced by two and three types of boundary. Figure 13-2 is the sister of the roadmap, showing the structural transformations that take us from one class of forms to another.

Reconstructing Brackets

Textual delimiters are intended to separate (or isolate) text from its context. They are used for grouping and for specifying order of operations. We have been using them in a quite different way, as objects and operators that define relative contexts.

Chapter 13

We begin with three fundamentally different ways to expose the containment imposed by textual delimiters: *enclosures*, *paths*, and *relations*. The Arrangement axiom provides some visual examples.

The natural form of brackets shows nesting. The top and bottom of each delimiter have been cut off in order to convert a two-dimensional enclosure into a pair of one-dimensional tokens.

$$(A\ [B\ C]) = (A\ [B])(A\ [C])$$

Reconnecting the fractured brackets generates the *enclosure dialects*. These dialects vary across the shape and dimension of the enclosures. For the three James boundaries, for example, we have been using three different shapes of enclosure. The **bounding box** (Figure 13-4) and **bucket** (Figure 13-5) dialects are examples of enclosure dialects. Enclosures make the contains relation explicitly visual as containment.

$$(A\ [\ _B\ _C\ ]) = (A\ [\ _B\ ])(A\ [\ _C\ ])$$

Alternatively, nested brackets can be extruded, identifying the levels of nesting as different tiers to generate the *relational dialects*. Each delimiting boundary is converted into a node in a network or into a territory in a map. The relational dialects exhibit containment as spatial connectivity. They are most prolific, leading to network and map varieties as well as to the physically manipulative dialect of **stacked blocks** (Figure 13-10).

$$\begin{matrix} & B\ C \\ & A[\quad] \\ (\quad &) \end{matrix} = \begin{matrix} B & & C \\ A[\ ] & & A[\ ] \\ (\quad)(& &) \end{matrix}$$

312

Dialects

The **network** dialect (Figure 13-6) connects boundary forms to the rich and well-studied domain of graphs and networks. The contains relation becomes a link between nodes. When nodes touch, links are turned into the shared borders of a map. The map dialects can maintain nesting information using a variety of visual and analogical cues. In the **map** dialect (Figure 13-7), nesting is indicated by overlapping, creating an iconic form that resembles stepping stones. The **wall** dialect (Figure 13-8) is a map variety that achieves nesting via an implicit gravity acting downward. The **room** dialect (Figure 13-9) maintains nesting through the availability of open doors that permit access to deeper rooms. Explicitly the outermost door leads outside.

$$\left(A\,[B\ C]\right) = \left(A\,[B]\right)\left(A\,[C]\right)$$

A third method of reconstructing fractured textual forms is to connect brackets to neighboring half-brackets. Kauffman calls these *capforms*.[7] This method constructs a single distinction between the objects within a bracket form by creating one interior and one exterior space that separates labeled forms into odd and even levels of nesting. The example of the Arrangement axiom below makes visually obvious that no forms have crossed their respective boundaries.

Capform connectivity generates the *path dialects*. A path dialect consists of *only one* instance of each type of boundary. Nesting in multiple enclosures is indicated by a path traveling across the singular border. The design of these dialects can get quite complex when there is more than one type of boundary, since different boundary types can support a greater variety of nesting relations. The **path** dialect (Figure 13-11) incorporates a *composite* single boundary that takes the place of a generic frame. Unlike the other dialects, paths cannot represent all possible James forms without occasional duplication of a boundary type, but they do illustrate the odd/even categorization of the location of variables. In our interpretation, the odd/even distinction is quite important since it separates exponential content from logarithmic content created by each round/square-bracket frame.

Selected Patterns

I've selected the three James axioms and the three important theorems from Chapter 5, in essence the core transformations of James algebra, to show in eight different spatial dialects. This selection of pattern equations is displayed in our typographical dialect in Figure 13-3.

13.3 Design

It is widely acknowledged that maintaining a formal semantics for diagrams is difficult. A boundary language with only one relation (depth-value and boundary logic are examples) makes formal diagrammatic structure more feasible. The diversity of spatial dialects permits designed nuances that can emphasize different aspects of container-based thinking. The eight dialects in this chapter — in addition to typographic strings — are certainly rough designs (and there are, of course, many more possible dialects). None have been empirically verified as beneficial and none have been independently demonstrated to be rigorous. All have a direct mapping to the textual delimiter representation of the James transformation patterns, and in that sense they are as formal as the textual delimiter dialect itself. The changes in perspective introduced by the sensual dialects are as structured as the symbolic manipulations that currently define rigor.

Dialects

([A]) = [(A)] = A	**inversion**
(A [B C]) = (A [B]) (A [C])	**arrangement**
A <A> = *void*	**reflection**
(A [ ]) = *void*	**dominion**
([A][o..$_N$..o]) = A..$_N$..A	**replication**
(A [<B>]) = <(A [B])>	**promotion**

Figure 13-3: *Pattern equations selected for display*

There are two central concepts that each dialect must represent unambiguously. One is *containment*, of course, showing which forms contain which other forms. The second concept is the *partial ordering* of containment relations, showing hierarchical nesting between forms and identifying an unambiguous outside.

All of the dialects default to the use of ellipsis, ..$_N$.., in Replication to indicate an arbitrary but finite number. Finite replication could have been wrapped up into a specialized variable or container. I've elected to let the ellipsis stand as an incursion into the form from the meta-language. It is a second-order function, a process that acts upon processes rather than forms. Alternatively we could say it is a variable about variables, not a form but an index of forms. These topics are better served by Volume II where we explore the structure of formalism.

Bounding Box Dialect

The bounding box dialect in Figure 13-4 is the mother tongue. Textual delimiters are derivative of a closed planar boundary that separates the plane into an inside and an outside. What's new are the three different types, or shapes, of bounding boxes, which could just as well be displayed as colors, roughness, width, labels, etc. Since we

Chapter 13

inversion

dominion

arrangement

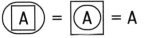

replication

reflection　　　　　**promotion**

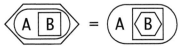

Figure 13-4: *Bounding box dialect*

can visually access a bounding box from-above (from outside of the plane of representation, from a third dimension), they are not experientially closed, as would be, for example, nested Russian dolls or rooms in a physical house.

Bucket Dialect

The bucket dialect (Figure 13-5) is a simple derivative of bounding boxes, this time with only the top of the container removed. Buckets are a conceptual metaphor for a generic physical container, so a bucket-like dialect elicits a different mental image than, say, a bounding box dialect. Images of buckets give more of a physical feel, we can imagine lifting and placing buckets one within the other.

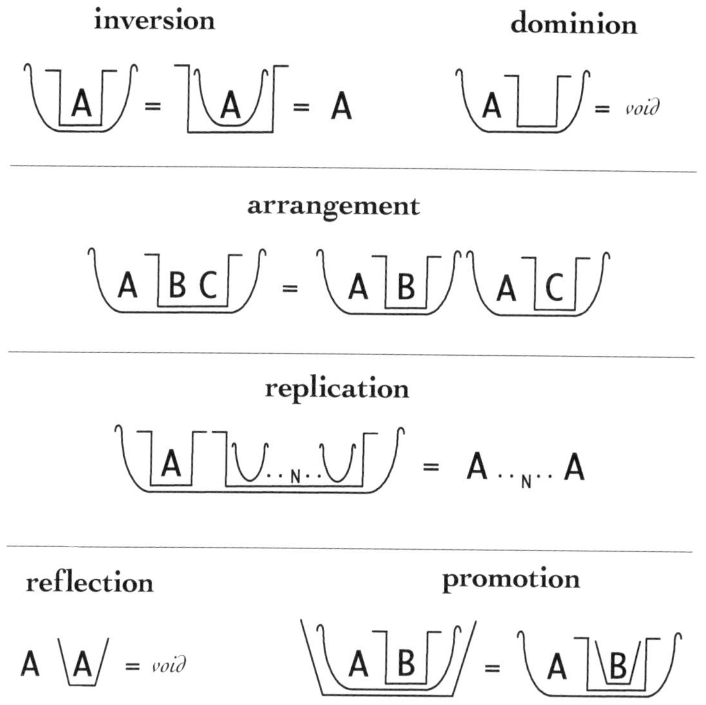

Dialects

inversion   **dominion**

$\lfloor A \rfloor = \lfloor A \rfloor = A$ $\lfloor A \quad \rfloor = void$

arrangement

$\lfloor A \lfloor B C \rfloor \rfloor = \lfloor A \lfloor B \rfloor \lfloor A \lfloor C \rfloor \rfloor$

replication

$\lfloor A \lfloor \cup .. N .. \cup \rfloor \rfloor = A .. N .. A$

reflection   **promotion**

$A \lfloor A \rfloor = void$ $\lfloor A \lfloor B \rfloor \rfloor = \lfloor A \lfloor B \rfloor \rfloor$

Figure 13-5: *Bucket dialect*

Buckets raise the question of the intent of the representation of a variable like A. Delimiters and enclosures give the feel that variables can stand alone, but a bucket leads us to wonder how to pick up and put a variable into a container. Each variable label could have been placed in a specialized variable-bucket, at the cost of adding visual clutter.

The bucket dialect is simply a rotation of the viewing perspective of bounding boxes through 90° out of the plane of the page. We are looking at the side rather than at the top. To be able to rotate implies access to a third dimension, so that the mapping from boundaries to buckets is inherently 3D. We have implicit access to an inside. The two-dimensional bucket dialect is a vertical slice through

Chapter 13

inversion   **dominion**

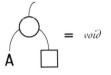

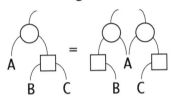

arrangement

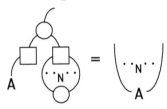

replication

reflection   **promotion**

 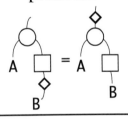

Figure 13-6: *Network dialect*

a collection of three dimensional, experiential objects. *Viewing perspective* (in this case from-above vs. from-the-side) plays an important role in subjective interpretation, and thus in learning and in application. The cost that conventional mathematics pays for choosing to separate image from meaning is both loss of expressibility and loss of comprehension.

Network Dialect

The network dialect (Figure 13-6) is certainly the most computationally tractable. It provides the image of nodes/locations/processors with links/paths/wires between them. Each link is a contains relation, so that the relational structure is explicit. Networks and other relational dialects have an additional display requirement to specify the direction of nesting, from shallowest to deepest, which requires directed links. The network version often presumes a gravitational metaphor, with deeper nesting shown at lower levels. For a network the level of nesting is defined by counting the nodes lying between beginning and goal (input and output).

Networks are a well-established modeling tool in computation and in mathematics. The structure of the generic frame is clearly visible in a network as a link between an upper round node and a lower square node. Multiple contents are multiple lower links. Deletion of structure is just disconnecting a link. Even the inverter diamond fits naturally into the patterns of flow. And networks support any number of types of container.

times

Unlike the enclosure dialects, networks do not require multiple replicas of variables. We can use a single node for each variable, and access the variable through multiple links, or pointers. The forms of Arrangement and Replication in Figure 13-6 show this feature clearly. The experience of driving on roads between different cities provides familiarity to the network approach of no replicated objects.

The fluidity of object and reference in networks can be expressed as a transformation rule that is unique to this (and similar) dialects, **structure sharing**. In structure sharing, nodes in a network that share the same linking structure can be joined into a single node with multiple links. Entire subnetworks are replaced by links to shared

Chapter 13

structure sharing

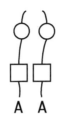   

structure. Structure sharing is not available in textual dialects. The textual representation of each of the above forms is ([A])([A]). Multiple occurrence of the same variable in a textual dialect emulates shared structure in a network dialect. The absence of structure sharing in symbolic mathematics has lead to rampant replication of symbols and a presumption that replication is free. As noted in Chapter 9, replication is the *source* of complexity.

Not only is notation not independent of meaning, it can also actively determine meaning. There is little structure sharing in the physical world since all physical objects are unique.[8] If two sheets of paper have the same sentence written on them, then the two *virtual* sentences share the same structure. However, the two *physical* sheets of paper do not share the same structure (i.e. they are not replicas) although they may share abstract properties such as use, color, shape, even content. Representational systems that lack structure sharing risk confusing features of the symbolic model with features of the physical circumstance being modeled.

Map Dialect

The map dialect (Figure 13-7) itself has many varieties. Maps use borders rather than links to identify containment. However, maps can be ambiguous about direction of nesting, about which territory is outermost. The roadmap in Figure 13-1 shows two large varieties of maps, distinguished by the location of the outermost container. The two varieties put the origin or entry either at an edge

Dialects

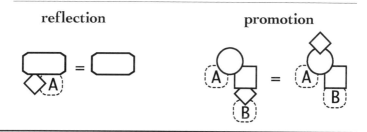

Figure 13-7: *Map dialect*

or in the center. Within each variety, the different maps emphasize different shapes for territories.

These map varieties have the outermost territory marked by a dot. We can interpret the dot as a *you-are-here* token, our place of entrance into the map. From our 3D perspective out of the plane of the page, we gain access to all territories equally, at the cost of losing our origin. In reading maps, pragmatically, we place a finger on the *I-am-here* point to identify the outermost starting point.

Chapter 13

containment as occlusion

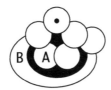 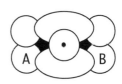

The map version shown in Figure 13-7 is of the **stepping stone** variety, where territories have a geometric shape and *occlusion of shape* indicates containment. The circular top stone is explicit as the only non-occluded shape. Occlusion can be interpreted as a gravitational metaphor, making territories into stairs, or into pools between waterfalls or into zones of influence. Variable labels cannot stand separately, they need their own outer shape in order to participate in an occlusion relationship.

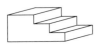

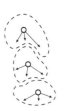

The Reflection rule has a structural cliff in the map dialect in Figure 13-7, an action at a distance if you will. There's a step down of two levels from the uppermost stone to the A stone. Reflection also requires the usually implicit outermost boundary to be explicit in order to establish that both the angle-territory and the A territory have a common neighbor in both the planar image and the three dimensional structure it represents. Notice also how the shared variable A in Arrangement changes its role from outlier on the left to a bridge on the right, from tangential on the left to essential on the right.

Wall Dialect

The wall dialect (Figure 13-8) is similar to the stepping stone dialect except oriented as a cross-section rather than from-above, the same distinction that differentiates the bounding box from the bucket dialects. The contained-by gradient is established gravitationally. The stone or territory directly below carries the weight of the stone above. Vertical boundaries are not shared, only

Dialects

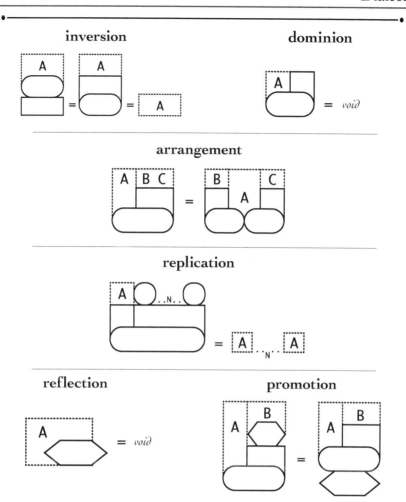

Figure 13-8: *Wall dialect*

horizontal boundaries. This dialect takes the ground, at the bottom of each figure, as the outermost container. In contrast to stepping-stones, Reflection does not need an explicit outer container since the ground is implicit. Like the map dialect, wall variables need spatial representation with explicit boundaries.

Chapter 13

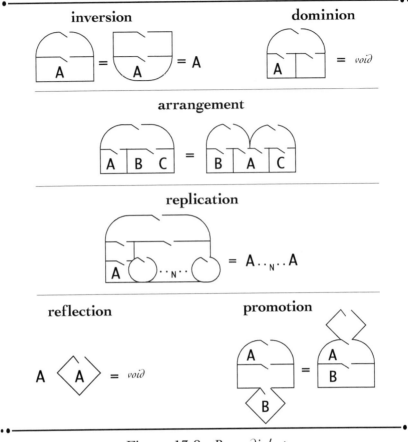

Figure 13-9: *Room dialect*

Room Dialect

The room dialect (Figure 13-9) bears a resemblance to the wall dialect. Here, none of the walls, vertical or horizontal, indicate containment. Rather a contained form has a door. The direction that the door opens specifies the nesting gradient. The shapes of rooms represent types of boundary. Variables occupy rooms, they do not need a room of their own. This design mixes the relational and the enclosure styles, since specific forms substituted for variables would need additional space. Structure sharing works only in some cases.

The room dialect is special because it represents an *experiential three-dimensional environment*. In Figure 13-9 we are looking down from-above into rooms, with the roof removed. The intent of this dialect is to provide an explorable space, where the structure of the form, and therefore the structure of arithmetic, can be experienced. Representation can be either an environment or an object. This dialect then has a version that exchanges the omniscient from-above perspective for a participatory from-within experience with limited knowledge of the surrounding structure or rooms. For example, in discovering the Dominion configuration, the participant might close the outer door permanently once the inner empty room has been located. Again Arrangement in the room dialect converts the variable A from *cul-de-sac* into nexus.

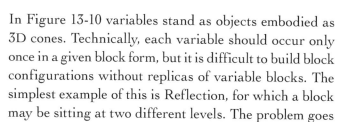

dominion

Block Dialect

The block dialect (Figure 13-10) is an extrusion of bounding boxes into a third dimension. Condensing dimension from-above converts blocks to bounding boxes. Condensing dimension from-the-side converts blocks to walls. Blocks are also intended to be experiential, but from an object-oriented rather than an environmental perspective. These arithmetic blocks can be manipulated, stacked and rearranged.

In Figure 13-10 variables stand as objects embodied as 3D cones. Technically, each variable should occur only once in a given block form, but it is difficult to build block configurations without replicas of variable blocks. The simplest example of this is Reflection, for which a block may be sitting at two different levels. The problem goes away, of course, if we allow flexibly shaped blocks, but

Chapter 13

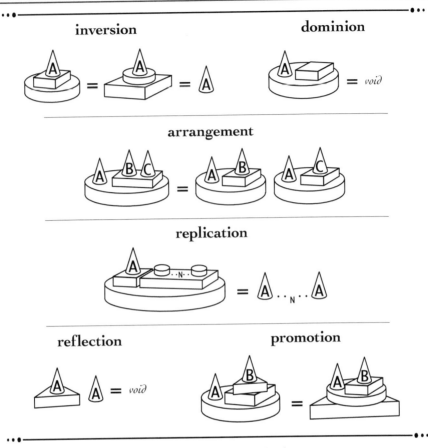

Figure 13-10: *Block dialect*

this would compromise the physical model of the dialect. To maintain physicality the variable A in Arrangement, for example, must be replicated. Although A could be placed as a bridge on the right-hand-side of Arrangement, this is not a general solution, as can be seen from the Replication theorem.

Path Dialect

The network and map dialects require only one reference to each variable. Taking that observation to an extreme, what would a form look like if we allowed only one

Dialects

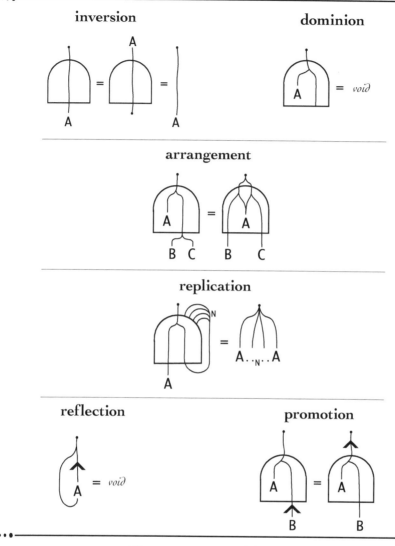

Figure 13-11: *Path dialect*

reference to each type of boundary? Based on capform connectivity, the path dialect (Figure 13-11) provides a single boundary for which the inside identifies odd depths of nesting and the outside identifies even depths. A path crosses over the same boundary several times to indicate deeper nestings.

In the path dialect, the outside dot indicates the starting point. The single boundary enclosure is composite, with a round boundary on top and a square boundary on bottom. This works well when dealing with inversion boundaries, since they usually occur as frames. A ground is indicated by a path that terminates on the boundary, such as in Dominion and Replication. Multiple sub-forms are represented by branching paths.

Think of variables as potentially available along an entire path. All variables are available along every path but are indicated only as terminals. In Arrangement, for example, the left side of the rule shows that variables A, B, and C all initially cross the round boundary. Variable A elects to stay within the round boundary, while B and C continue on to cross the square boundary border. There they part ways. On the right side of the rule, the path splits first on the outside, creating two crossings of the round boundary. Variable A reunites with itself once inside. B and C continue separately across the square boundary.

Here is an example of the rubber-sheet conversion of the network dialect to the path dialect.

converting a network to a path

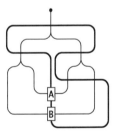 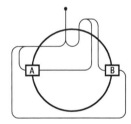

Paths work well when there is only one type of boundary. When there is more than one boundary type, paths do not provide full expressibility. Multiple types of boundary can be arranged in multiple ways so that the outside/inside distinction cannot be maintained on a flat surface. This situation is similar to a planar network display that has links that must cross in two-dimensions. For example, forms like <[(<A>)<B>]> require replicating the angle boundary. For the specific rules in Figure 13-11 this problem is not severe.

Like the room dialect, the path dialect is potentially experiential. We can be inside the representation, walking along the path with local knowledge of our location, or, as is necessary in print, we can sit in a higher dimension observing the entire form from-above with complete knowledge. Path forms are also compatible with the 3D manipulative perspective. Variables might be tennis balls, for example, while paths are valleys along which variables can roll. Paths could be streams with variables that are fish swimming in them. Paths could be, well, paths in the forest with variables as animals roaming on them.

13.4 Applications

For these sensual dialects, the transformation rules standing alone are not motivating. The purpose of transformation is to calculate, demonstrate and derive. We'll show the James postsymbolic forms in action for one member of each class of dialects:

— enclosure (bounding box)
— relation (network)
— path (path)

For content, we'll show three different types of application:

— calculation of 5 ÷ 2
— derivation of the Dominion theorem
— demonstration that $\log_{1/B} N = \log_B (1/N)$

The sensory dialects can also be animated, providing a smoothly flowing, dynamic display of arithmetic in action. Void transformation, for example, could be shown as elements fading into the background, or burning up, or growing legs and walking away. Structural rearrangement could be animated by elements merging, diverging, expanding, or growing legs and moving to a different location. Some animation techniques such as fading away to represent void-equivalence are strongly connected to the transformational intentions of the calculus itself, while some are fanciful add-ons that may entertain while distracting from the mathematical formalism. Here is where rigor and psychology intersect.

Calculation

Figure 13-12 shows a calculation, 5 ÷ 2, in the bounding box, network and path dialects. The three models bear a strong resemblance. Each illustrates the same sequence of transformations in the same order, only the outward appearance differs. From the perspective of animation, the image sequences are frames stopped at roughly the same time.

The textual representation of the example calculation is

```
                5 ÷ 2 ☞
                ([ooooo]<[oo]>)
disperse        ([oo]<[oo]>)([oo]<[oo]>)([o]<[oo]>)
cancel          (           )(           )([o]<[oo]>)
clarify         (           )(           )(    <[oo]>)
                                              ☞ 2 ½
```

What is missing from the stepwise display is available concurrency. The textual form above requires discrete steps, however the second cancel step and the third clarify step could have taken place at the same time. If we count "steps" as rule applications, then any theorem hides the potentially many steps that comprise it. Therefore only the application of axioms can be considered as countable transformation steps. It is possible, for example, to name the expression 5 ÷ 2 directly as a theorem that takes only one step. Similarly, transformations that apply at the same time can be counted as single steps, the theorem spanning time rather than space. The net result

Dialects

BOUNDING BOX DIALECT

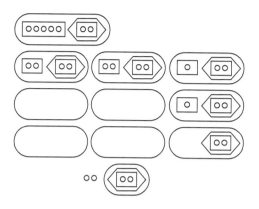

NETWORK DIALECT

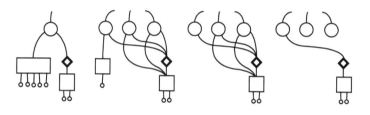

PATH DIALECT

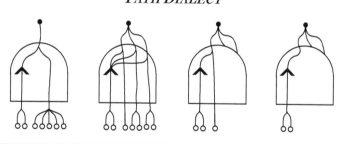

Figure 13-12: *Calculation of 5÷2 in three dialects*

of these observations is that the presentation of conventional calculations and proofs is constrained by the needs of communication, essentially an appeal to psychology rather than to rigor. What we think of as computational or deductive *steps* are artifacts of notation, perspective, cognitive preference and implementation.

Chapter 13

BOUNDING BOX

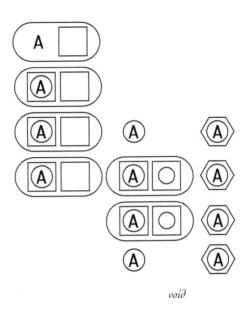

Figure 13-13: *Bounding box derivation of Dominion*

Note the structure sharing in both the network and the path dialects in Figure 13-12. This economy of form is not available in the bounding box or textual dialects. The figure does stop short of complete structure sharing since the units themselves retain their individual representations. The network representation could have incorporated multiple links to a single *unit* node. The path could have doubled back with N branches terminating when they reach the single round boundary.

structure sharing of units

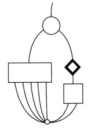
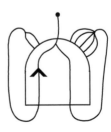

Dialects

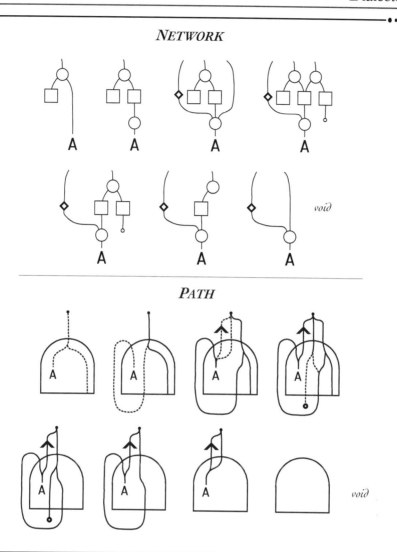

Figure 13-14: *Network and path derivations of Dominion*

Derivation

Figure 13-13 shows a derivation of the Dominion theorem, expressed in the bounding box dialect while Figure 13-14 shows the same derivation in the network and path dialects. The textual transformation sequence uses all three axioms, although two applications of Inversion

Chapter 13

are combined into the Indication theorem (the mark and unmark transformations).

	(A [ ])	
enfold	([(A)] [ ])	
create	([(A)] [ ]) (A)	<(A)>
mark	([(A)] [ ]) ([(A)][o])	<(A)>
collect	([(A)][o])	<(A)>
unmark	(A)	<(A)>
cancel	*void*	

The bounding box derivation follows closely that of the textual version above. However, the network derivation in Figure 13-14 incorporates structure sharing. A consequence is that the conceptualization of what Dominion means and how it is derived differ significantly. The hybrid textual form below illustrates the impact of structure sharing on methodology. The textual technique requires dummy variables to name shared structure. Since variable labels occur more than once, this might more accurately be called label sharing.

	(A [ ])	
enfold	([(A)] [ ])	
create	([(A)] [ ]) (A)	<(A)>
substitute S=(A)	([S] [ ]) S	< S >
mark	([S] [ ]) ([S][o])< S >	
substitute T=[S]	(T [ ]) (T [o])< S >	
collect	(T [o])	<S>
clarify	(T)	<S>
substitute	([S])	<S>
clarify	S	<S>
cancel	*void*	

The first two steps are the same. Structure sharing at the third step introduces via substitute the newly named variable S. At the fourth step (the top right network form) marking S creates another named variable T. It is T that serves as the frame type for the application of collect. Now we need to step back to rectify the labels by returning, via substitution, T to [S]. It is this **namespace management** that creates great difficulties for computational systems. It's why our URLs are so unwieldy. This example also illustrates that the free generation of names is not computationally free.

The path derivation of Dominion in Figure 13-14 highlights changes in the path at each step. The path dialect enforces structure sharing due to its minimality of structural reference. Paths are explicitly processes.

Demonstration

The derivation of the Dominion theorem stays within the James system, so we'll also show an example of rules and structures within conventional algebra. Let's look at the structure of a logarithm with a unit fraction base. Chapter 12 has a textual demonstration of this equality.

$$\log_{1/B} N = \log_B (1/N)$$

Expressed in exponential form,

$$\log_{1/B} N = R \quad \Leftrightarrow \quad \begin{aligned} (1/B)^R &= N \\ 1/B^R &= N \\ B^R &= 1/N \end{aligned}$$

Now we can take the $\log_B$ of both sides,

$$\log_B B^R = \log_B (1/N)$$
$$R = \log_{1/B} N = \log_B (1/N)$$

Figure 13-15 shows the James forms of this demonstration in the bounding box dialect; Figure 13-16, in the network dialect; and Figure 13-17, in the path dialect. For comparison, the textual delimiter form of the transformation follows.

Chapter 13

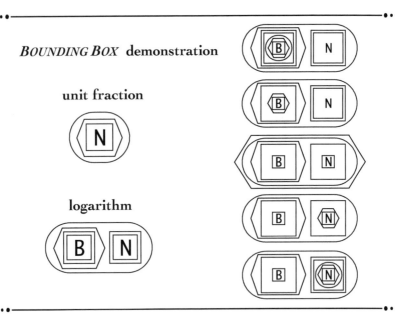

Figure 13-15: *Bounding box demonstration of unit-fraction bases*

Recall the form of a unit fraction and the form of a logarithm.

$$1/B \quad ☞ \quad (<[B]>)$$
$$\log_B N \quad ☞ \quad (<[[B]]>[[N]])$$

$\log_{1/B} N \quad ☞$

substitute	(<[[(<[B]>)]]>[	[N]	])		
clarify	(<[	<[B]>	]>[	[N]	])
promote	<(<[	[B]	]>[	[N]	])> ☞ $-\log_B N$
demote	(<[	[B]	]>[	<[N]>	])
enfold	(<[	[B]	]>[[(<[N]>)]]) ☞ $\log_B 1/N$		

Certainly the above textual bracket demonstration looks more complicated than the simple transformations of exponents and logarithms in conventional notation. The bounding box demonstration in Figure 13-15 also looks complicated. However there is a significant difference. The James derivation calls upon one theorem (Promotion) and the three James axioms. If we count all

Dialects

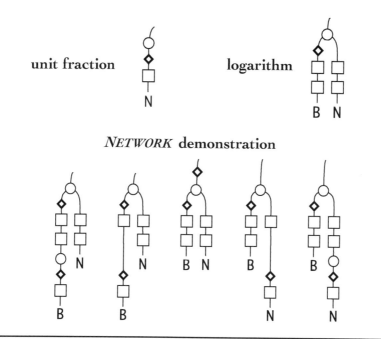

Figure 13-16: *Network demonstration of unit-fraction bases*

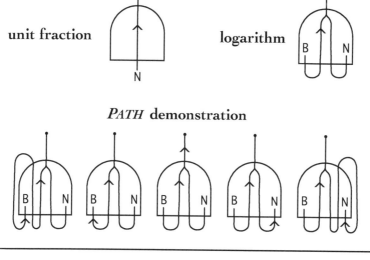

Figure 13-17: *Path demonstration of unit-fraction bases*

applications of the three axioms required for this demonstration, including those used to demonstrate theorems like Promotion, it takes about a dozen transformations at the atomic level to achieve the demonstration. In comparison, if we look closely, the conventional derivation calls upon nearly the entire axiom structure of exponentials and logarithms. In particular

$\log_M N = R \Leftrightarrow M^R = N$ *definition of logs as exponents*

$(1/M)^R = 1^R/M^R$ *distribution of exponents*

$1^R = 1$ *power of 1*

$\log_M M^R = R \log_M M$ *log of powers*

$\log_M M = 1$ *log of base*

To get down to the grain-size of a formal proof comparable to the James version, each of these rules would need to be demonstrated from axioms and definitions. Two require inductive proof, which itself introduces the theory of Induction. The logarithmic rules are built upon exponential rules which are built upon addition and multiplication rules which are built up the definition of number which are all built upon set theory which is built upon predicate calculus. That's over two dozen axioms alone.

However, the main point here is that the figures show the demonstration in visceral terms, as rearranging physical blocks (with our view of the blocks shown from-above), as rearranging nodes in a tree (a network without re-entry or structure sharing), and as traveling along a path passing signposts (that mark boundary crossings). Each shows clearly the symmetries originally in the unit fractions as base and argument. I find particularly appealing the path perspective on the structure of the logarithm and on the symmetric transformations that shift the left-most path of B in the first step to the right-most path of N in the last step.

13.5 Remarks

The figures in this chapter show the unfolding of numeric computation in fine grain steps, from fundamental principles. They demonstrate that rational arithmetic can be presented in many diagrammatic and sensory dialects. They also expose underlying structural symmetries and dependencies that are hidden in our current notations. No, these techniques are not intended to teach school children, nor adults for that matter. They are intended to put a rigorous postsymbolic foundation underneath arithmetic, a foundation that is much less complex than Peano axioms (which define only addition and multiplication), than inductive principles (which require excursions into infinity), and than the group theoretic assumptions (which enforce irrelevant axioms) underlying our current conception of simple arithmetic. Importantly, the different visual and visceral dialects encourage substantively different cognitive models, different multimodal ways to understand arithmetic.

Verification techniques for diagrammatic formalisms are in their infancy for two reasons. The obvious reason is that over the last century formalism has been defined to apply *only* to strings of symbols. It is an open question whether the current definition of formalism itself is adequate. The second reason is that only a very few researchers are interested in formal foundations. Most are content to believe what was established a century ago. The exception is the relatively young Computer Science community, who have actively extended the realm of rigor to include efficiency. The challenge of verification of correctness for silicon processors is not whether verification is possible, but rather what size systems can be verified in a finite and tractable amount of time.

Now we'll take a brief look at some of the precursors to the work in this volume.

Chapter 13

Endnotes

1. **opening quote:** B. Rotman (2000) *Mathematics as Sign: writing imagining counting* p.68.

2. **a sensory exploration of postsymbolic forms of James algebra:** For the visually-oriented much of the narrative is gratuitous. Then again, that has been the case throughout the entire volume.

3. **the representation of formal concepts interacts strongly with how we think:** The Sapir-Whorf linguistic hypothesis posits that the structure of our language strongly influences the cognitive and behavioral options exhibited by a culture. The *Participation Principle* is an application of Sapir-Whorf to formal systems. Rigorous thinking is partially defined by the tools of thought.

4. **not analogous to different spoken languages, such as English, Latin, and Farsi:** Symbolic mathematics emulates the spoken word; both are temporal strings of encrypted signals, one encoded in sound, the other in shape. After digital convergence, multimodal encoding is just another way to specify which *output device* is in use, whether speaker or pixel display or touch-sensitive screen or subcutaneous metabolism monitor.

5. **pictorial arguments, while no so purely conventional, can be quite legitimate:** J. Littlewood (1986) B. Bollobás (ed.) *Littlewood's Miscellany* p.54.

6. **make anything other than symbolic notation prohibitively cumbersome to use:** K. Devlin (2000) *The Math Gene* p.8.

7. **Kauffman calls these capforms:** L. Kauffman (1993) *Knots and Physics* 2nd ed. p.605-606.

> ... [capforms are] a combinatorial algebra on parenthesis structures — the boundary logic — that forms a foundation for the Temperley Lieb algebra.

8. **since all physical objects are unique:** *Structure sharing* is different than structural abstraction. Sharing joins together multiple references to the same form, while abstraction ignores selected features. Human prejudice probably does not apply to structure sharing since sharing is a property of representation rather than reality. We do have patent laws that adjudicate cases of structure sharing in creative works.

Chapter 14

Alternatives

Laws of Form is about the power of nothing. This is difficult to acknowledge for many persons, that it all comes from nothing. That what is, what exists is limited, bounded and identical with what is not.[1]
— *Louis Kauffman (2011)*

The boundary math community has developed several versions of iconic arithmetic, each based upon different rules (structural invariants) for containment relations. The briefly described systems in this chapter fall into two main categories: simpler versions of James algebra and variations on depth-value ensembles.

Spencer Brown numbers came first. His original work inspired and guided all other systems described in this volume. The common thread of Spencer Brown and James-like systems is multiplication based on an Arrangement rule. Kauffman string numbers came first to inspire and guide development of depth-value systems. Their common thread is multiplication based on substitution. All systems incorporate and depend upon the Additive Principle.

In this chapter, we'll look at Spencer Brown numbers and the improvements designed by Kauffman; at Kauffman two-boundary arithmetic which sits neatly between the single boundary forms of Spencer Brown and the three boundary forms of James; and at Kauffman string arithmetic. We'll also explore Spatial Algebra, which incorporates a new type of multiplication. We'll examine models of multiplication as the primary difference between all these additive arithmetics:

— structural multiplication implemented by an Arrangement rule (Spencer Brown, Kauffman two-boundary, James)

— multiplication-as-substitution (Kauffman string, depth-value)

— multiplication as touching (spatial algebra)

Conventional multiplication tables and the common grade-school multiplication techniques are respectively rote and algorithmic applications of multiplication-as-substitution. Matrix multiplication and the Cartesian product of relational calculus are structural.

14.1 Spencer Brown Numbers

Spencer Brown applied his pioneering perspective on boundary logic to the development a boundary algebra of numbers in 1961.[2] Spencer Brown numbers use only one type of boundary, they achieve diversity by incorporating two different types of context, additive and exponential. Since Spencer Brown, together with C. S. Peirce, invented boundary logic, his notation for a boundary delimiter is in wide use among the *Laws of Form* community. The **Spencer Brown cross** is a boundary delimiter that looks like this:

$$\rceil$$

It is a clever variant of the negation sign, ¬, extended to have an inside space. It is also an abbreviation for a square container, □. Although there is a strong rationale for continuing to use the Spencer Brown cross here, for

1-elements		0-elements	
0	☞	0^0 ☞	( )
1	☞ ( )	0^1 ☞	(( ))
2	☞ ( )( )	0^2 ☞	(( )( ))
3	☞ ( )( )( )	0^3 ☞	(( )( )( ))

Figure 14-1: *Elements within Spencer Brown numbers*

convenience and consistency I will defer to delimiting typographical brackets and to diagrammatic forms.

Arithmetic

Figure 14-1 shows the elements of the Spencer Brown arithmetic. There are two types of element, corresponding to the additive null concept of 0 and the multiplicative null concept of 1. To **cross** a form is to put it inside a boundary. In general crossing an element changes its type. Thus, the type of an element is determined by its level of nesting. Forms nested at even depths are **1-elements** (i.e. units) while forms nested at odd depths are **0-elements** which are interpreted as exponents (i.e. powers of the *context* of the form).

Spencer Brown and James algebras have the same structural definitions of addition and multiplication, with the exception that Spencer Brown forms consist of only one variety of container.

$$A + B \quad ☞ \quad A\ B$$
$$A \times B \quad ☞ \quad ((A)(B))$$
$$B^A \quad ☞ \quad B\ (A)$$

Spencer Brown is clear that forms do not have a relationship across space, so that addition, for example, includes an implicit outermost bracket.

Chapter 14

The only relationship between elements is continence; a given element is or is not contained in another given element.[3]

Projecting a relationship across "shared" space, such as

A B ☞ same-space[A, B]

is the dominant error that folks make[4] when trying to understand the structure of boundaries. This point is important enough to emphasize. In boundary math systems, *there is only one relation*. This difference alone is sufficient to distinguish boundary numbers from conventional set-based group theoretic numbers. This from Wikipedia:

> In mathematics, a ring is one of the fundamental algebraic structures used in abstract algebra. It consists of a set equipped with two binary operations that generalize the arithmetic operations of addition and multiplication.[5]

How can we have an arithmetic with only one relation? It is tempting to say that addition is absorbed into shared space, but this is not the case, since space has no properties. In Spencer Brown numbers, addition is multiple forms contained by the same container.

A + B ☞ (A B)
☞ ( ) contains A, ( ) contains B

The shell-brackets here are our generic value-free outermost container. Multiplication is containment two levels deep.

A x B ☞ ((A)(B))
☞ ( ) contains (A), ( ) contains (B)

Thus the distinction between two binary operators is converted into a distinction between levels of nesting within one relation.

AXIOMS

(( )) = *void*	**universe**
((A)(B)...) C = ((A C)(B C)...)	**transfer**

THEOREMS

((A)) = A	**cancellation**[6]
A ( ) = ( )	**null power**

INTERPRETATION

(A B)	☞	A + B	**addition**
((A)(B))	☞	A × B	**multiplication**
(B (A))	☞	B^A	**exponentiation**

Figure 14-2: *Structure of Spencer Brown numbers*

Kauffman makes the observation that in the Spencer Brown system, N is a number, while (N) is the operator associated with that number. The outer boundary distinguishes passive from active, description from operation. There is some elegant boundary thinking here. From the inside, we see multiplicity. Moving to the outside we see a bounded multiplicity that can be accessed and manipulated by its boundary.

Axioms

The two axioms of Spencer Brown numbers are presented in Figure 14-2. These axioms correspond, roughly, to Inversion and Arrangement in the James system. In fact, Spencer Brown proves Inversion as a theorem.

((A))	
(( )) A	transfer
A	universe

Chapter 14

Here is an example of the structure of the form of exponentiation.

$$A \times A \times A \times A \quad \Rightarrow \quad ((A)(A)(A)(A))$$
transfer
$$((\ )(\ )(\ )(\ ))\ A$$
hybrid
$$(\quad 4 \quad)\ A \quad \Rightarrow \quad A^4$$

And here's the demonstration of a conventional rule for exponents.

$$A^{(B+C)} \quad \Rightarrow \quad A\ (\quad B \quad C\ )$$
cancel
$$A\ ((\ (B))(\ (C)))$$
transfer
$$((A\ (B))(A\ (C))) \quad \Rightarrow \quad A^B \times A^C$$

Spencer Brown proves the rules of exponents, and in his short appendix on numbers[7] states that the seven standard rules of common algebra as defined by group theory also have simple proofs.

The dominant motivation for Spencer Brown numbers is to construct both logic and numbers from the *same* containment patterns. Unit Accumulation (Chapter 9) is the foundation for the generation of natural numbers. When Unit Accumulation is denied, ( )( ) = ( ), we get an idempotency rule that is the foundation of Spencer Brown's boundary approach to logic. The four transformation rules in Figure 14-2 remain the same for both logic and arithmetic. Thus Spencer Brown has demonstrated the common roots shared by both logic and numerics. Jack Engstrom has extended this communality to include finite set theory.[8]

Issues

There are some problems with the purity of Spencer Brown numbers that both Lou Kauffman and Jack Engstrom have addressed.[9] For example there is an interpretive ambiguity concerning the model of exponentiation. Exponents and addition are both interpretations of sharing the same container.

(N) A ☞ A^N

(N) A ☞ $0^N + A$

Spencer Brown introduces an order of operations to resolve this problem. But in a boundary math, the boundaries themselves should indicate grouping and precedence.

Another issue is that the structure of transfer must be limited to specific forms. Given

A ((B)(C)) = ((A B)(A C))

the forms not being transferred, the B and C that remain in one location, must be of the same element type. An additional constraint is that A, B and C cannot all three be 1-elements. These restrictions are equivalent to extending the pattern rules into many different sub-rules. Alternatively the restrictions create two different classes of variables to stand in place of the two types of elements. Here's what transfer would look like to a pattern-matcher. Subscripts indicate the element type.

A_0 ((B_0)(C_0)) = ((A_0 B_0)(A_0 C_0))
A_0 ((B_1)(C_1)) = ((A_0 B_1)(A_0 C_1))
A_1 ((B_0)(C_0)) = ((A_1 B_0)(A_1 C_0))

There are other cases for which an intermediate form generated during transformation has an ambiguous interpretation. It is most desirable that every transformation step maintain the integrity of the interpretation. For example, the form of exponentiation is B (A), where B is the base and A is the exponent. How should the form (A) standing alone be interpreted? Engstrom makes the case that it has no numeric interpretation.[10]

The entire system of constraints, together with the 0- and 1-element categories, can also be seen as limiting the application of axioms to various odd and even depths in a nested expression. Odd and even depths in turn can be expressed by using two different types of brackets.

Chapter 14

	NUMBERS		INTERPRETATION
0	☞ ( )		
1	☞ (( ))	((A)(B))	☞ A + B
2	☞ (( )( ))	A B	☞ A × B
3	☞ (( )( )( ))		
N	☞ (( )..$_N$..( ))		

AXIOMS

((A)) = A *all contexts*

( )( ) = ( ) *multiplicative context*

A ((B)(C)) = ((A B)(A C)) A *in multiplicative context*

Figure 14-3: *Kauffman numbers in the form*

Kauffman Numbers in the Form

Lou Kauffman has been a leader in exploring the formal mathematics of Spencer Brown's innovations. In his monograph *Arithmetic in the Form* Kauffman describes a dual to the Spencer Brown numbers that resolves their interpretative ambiguity.[11]

Kauffman's version shares the Additive Principle with both Spencer Brown and James numbers, although addition involves a transformation that resembles fusion. Here's Kauffman:

> A number is a container filled with "that many" objects (marks). To add two numbers, cancel their containers and put the contents together in a single container.[12]

Figure 14-3 shows the structure and rules of Kauffman's system. Integers are bounded and correspond to Spencer Brown's 0-elements. They form a tally system with an explicit outer boundary. Because Kauffman's system has only one type of boundary, the outer boundary is not different than that of the unit boundaries. The form of addition is also the form of a number. Multiplication is juxtaposition (sharing the same container). Thus

$$(( )( )) \; ☞ \; 2$$
$$A \; (( )( )) \; ☞ \; 2 \times A$$
$$((A)(A)) \; ☞ \; A + A$$

Here's an example of addition at work. The Cancellation rule manages reduction of the outermost boundaries of individual numbers. Below, the form of addition is initially highlighted.

$$2 + 3 \; ☞ \; ((\; (( )( )) \;) \; (\; (( )( )( )) \;))$$
$$(\quad ( )( ) \qquad ( )( )( ) \quad) \; ☞ \; 5 \quad \text{cancel}$$

To address the ambiguity found in Spencer Brown numbers, Kauffman makes a *contextual* restriction that numbers must have an outermost container, so that juxtaposed forms are not numbers, but rather multiplication operations that are enacted using Transfer. For example, $2 \times (1 + 3)$ is

$2\,((1)(3))$ *with* $1=(( ))$, $2=(( )( ))$, $3=(( )( )( ))$

$(( )( )) \; ((( ))) ((( )( )( )))$	substitute
$(( )( )) \; (\quad ( ) \quad ( )( )( ) \quad)$	cancel
$(((( )( ))) \; ((( )( ))) \; ((( )( ))) \; ((( )( ))))$	transfer
$(\quad ( )( ) \quad ( )( ) \quad ( )( ) \quad ( )( ) \quad)$	cancel
$\qquad\qquad\qquad ☞ \; 8$	

We must suppress reading the interior of a number as a multiplication. In particular

$$(( )( )) \; ☞ \; 2$$
$$( )( ) \; ☞ \; 0 \times 0$$

Multiplying by zero requires a specialized contextual rule, what we have identified in Chapter 7 as Unify.

$$( )( ) = ( )$$

This rule applies only when an outermost bracket is missing. One way to describe this type of constraint is to differentiate between levels of nesting. Kauffman identifies alternating additive and multiplicative levels.[13]

additive context = (*multiplicative content*)
multiplicative context = (*additive content*)

This distinction is analogous to Spencer Brown's 0-elements and 1-elements. Rules trigger only in specific alternating nesting levels. This contextual finesse is equivalent to using two types of boundaries (round and square) together with adding an Inversion rule to enforce alternation.

14.2 Kauffman Two Boundary Form

Kauffman explores a two boundary version of Spencer Brown numbers that is shown in Figure 14-4.

Round brackets denote a boundary whose inside is additive and square brackets indicate a boundary whose inside is multiplicative.[14]

The two boundaries define two values of a unit.

( ) ☞ 0
[ ] ☞ 1

The Cancellation rule takes the familiar form of Inversion.

cancellation   ([A]) = [(A)] = A

The interpretive dilemma remains. When units are all of the same type, i.e. ( )( )( ) or [ ][ ][ ] for example, then computation can go forward. The former is a multiplication of zeros, the latter is a multiplication of ones. However ( )[ ] violates the syntactic rules of the system, and is not interpreted as 0 x 1. "...this does not imply that ( )[ ] has a value because this expression is not arithmetical."[15] The interpretative ambiguity has been transferred to a syntactic question of the numeric legitimacy of specific forms.

Alternatives

	NUMBERS		INTERPRETATION		
0	☞ ( )	((A)(B))	☞	A + B
1	☞ [ ]	A B	☞	A x B
N	☞ (( )..ₙ..( ))	[(A)..ₙ..(A)]	☞	A^N

A *and* N *in different contexts*

AXIOMS

[(A)] = ([A]) = A

A ((B)(C)) = ((A B)(A C))

Figure 14-4: *Kauffman two-boundary arithmetic*

Exponentiation takes on the form suggested by Spencer Brown, but without the ambiguities. The interpretative restriction is embedded into the level of nesting. "By definition, a^b is defined only if a and b are in opposite universes."[16] The universes, of course, are the two different odd or even levels.

Kauffman mentions a derivable special case,

$A^{()}$ = [ ] ☞ A^0 = 1   hybrid

which also holds for A = ( ). That is, 0^0 = 1. This conventionally controversial definition now becomes a structural conclusion.

Difference

Kauffman remarks:

> The whole purpose of this construction of formal arithmetic has been to show that ordinary arithmetic is really quite extraordinary -- that it grows out of our abilities to remember and repeat and our abilities to form patterns and contexts.[17]

Chapter 14

Kauffman directly identifies the revolutionary core ideas in the study of boundary numbers.

- All numeric form is built from *void*.
- There is no difference between numbers and operations.
- Arithmetic arises directly from logic by converting unification into accumulation.

We have thus far seen three different methods that differentiate addition from multiplication.

- different types of elements (Spencer Brown)
- alternating levels of nesting (Kauffman)
- different types of boundaries (James, Kauffman)

Kauffman's work has made clear that these methods are all equivalent but for notation.

Spencer Brown's model of arithmetic-as-distinctions shows that our culturally universal arithmetic is baroquely complicated, making distinctions where none exist and confounding syntactic bookkeeping with numeric concept. Kauffman again:

> Numbers are not collections that can be combined via the properties of addition and multiplication. Rather, numbers are the residues of the equivalences of forms under the processes of crossing, calling, juxtaposition, and recursion, the true precursors of addition and multiplication. The numbers occur inextricably mixed with their own operative powers.[18]

We'll now move from the precursors and variants of James algebra to the precursors and variants of depth-value arithmetic. A continuing theme is the comparison of two models of multiplication.

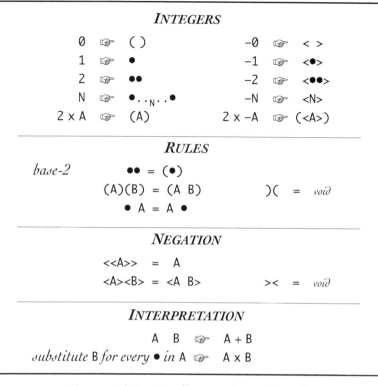

Figure 14-5: *Kauffman string arithmetic*

14.3 Kauffman String Arithmetic

In *Arithmetic in the Form*, Kauffman includes an appendix of earlier work on a string arithmetic that is the direct precursor of the depth-value arithmetic described in Chapter 3.[19] This arithmetic is presented in Figure 14-5. You will recognize it as depth-value arithmetic without parallelism. • is a unit, the same as unit-ensembles. Kauffman provides base-2 examples; the grouping rule converts two units into one nested unit.

Historically, as work on boundary mathematics proceeded over the last half-century, many authors have used a diversity of typographical delimiters to represent the operations of arithmetic, without much consistency. String arithmetic was an early system, in which

Chapter 14

INTEGERS

```
0  ☞  ( )
1  ☞  •                    -1  ☞  •̄
2  ☞  •• = (•)             -2  ☞  •̄•̄ = (•̄)
N  ☞  •..ₙ..•              -N  ☞  •̄..ₙ..•̄
```

RULES

$$• \; \bar{•} = void$$
$$•)\bar{•} =)•$$
$$\bar{•})• =)\bar{•}$$

Figure 14-6: *Kauffman string arithmetic with negative units*

Kauffman used the angle bracket as the depth discriminator for the Group operation, and the square bracket as negation. Purely in the interest of readability, I've altered Kauffman's original notation to be consistent with the other systems in this volume. Specifically, for Kauffman's original string notation,

— the grouping bracket < > becomes ( )
— the negation bracket [ ] becomes < >

Both grouping and negation can be formulated as void-based operations. Louis Kauffman explains:

> The brackets take the roles of place markers and "carry operators." Carrying happens automatically when place-marker boundaries collapse through the rule)(= *void*.[20]

The string rearrangement rule, gives permission to move units along the string representation, in order to form adjacent units that trigger the grouping rule.

$$• \; A = A \; •$$

For example

$$•(•)• = (•)•• = (•)(•)$$

Subtraction

Like the James angle-bracket, in Figure 14-5 negative forms are placed within a negating boundary. Here is how the substitution rules handle multiplication of negative numbers.[21]

$$\bullet \ \times \ <\bullet> \ \ ☞ \ \ (\!(<\bullet> \bullet \ \bullet \ )\!) \ \ \Rightarrow \ <\bullet>$$

$$<\bullet> \times \ \bullet \ \ ☞ \ \ (\!(\ \bullet \ \ \bullet \ <\bullet>)\!) \ \ \Rightarrow \ <\bullet>$$

$$<\bullet> \times <\bullet> \ \ ☞ \ \ (\!(<\bullet> \bullet <\bullet>)\!) \ \ \Rightarrow \ <<\bullet>> = \bullet$$

This is the same case as the use of negative ensembles such as <●●●> in Chapter 2. Not only does Kauffman subtraction include Reflection as a natural consequence, the angle-bracket theorems are the same for both systems.

Kauffman has also developed a version of his string arithmetic that incorporates subtraction using a **negative unit**, $\overline{\bullet}$, similar to the unit-ensemble ◊ of Chapter 2. Figure 14-6 shows this system. Here's a *base-2* example.

$$8-1 \ \ ☞ \ \ (((\bullet \ ) \ ) \ )\overline{\bullet}$$
$$(((\bullet \ ) \ )\overline{\bullet})\bullet$$
$$(((\bullet \ )\overline{\bullet})\bullet)\bullet$$
$$(((\bullet\overline{\bullet})\bullet)\bullet)\bullet$$
$$(((\ \ )\bullet)\bullet)\bullet$$
$$((\ \ \ \ \bullet)\bullet)\bullet \ \ ☞ \ \ 7$$

The boundary crossing rules are particularly interesting because they are analogous to carrying and borrowing in place-value notation (*base-2*).

14.4 Spatial Algebra

One of the objectives of spatial algebra is to identify the natural, intuitive concepts necessary for introductory algebra, and to contrast these concepts with the perspectives introduced by the abstract algebra taught in today's schools. Spatial algebra is a learning tool for those first encountering algebra.[22] There is, however, a significant conceptual change. The concepts associated with space are different than those associated with conventional algebraic structures such as groups. From an intuitive perspective, operations embedded in space apply to all objects in that space. By reintroducing the Additive Principle into algebra, the concepts of algebra change. They become simpler. The Participation Principle allows students to call upon their physical and perceptual knowledge and experience.

Spatial algebra uses the three dimensions of natural space to express algebraic concepts. Numbers and variables are represented by physical blocks.[23] Figure 14-7 shows some examples of the structures of spatial algebra. Blocks labeled with digits presume the digit facts for addition and multiplication. The aspects of three-dimensional space that are used for the representation of algebraic concepts include:

- empty space (the void)
- partitions between spaces
- labeled objects that share a space
- labeled objects that share a boundary.

The Additive Principle still holds, addition is placing blocks within the same (implicitly bounded) space. Multiplication, in contrast, is simply blocks that touch. Contact is not containment because blocks touch symmetrically. Block A touches Block B and concurrently Block B touches Block A. The touches relation is an equivalence relation since touches is **reflexive** (a block touches itself), **symmetric** (blocks mutually touch) and **transitive** (touching blocks all touch).

The implicit outermost container is the ground upon which blocks sit. The gravitational metaphor is not part of this representation system, but it is convenient to think of blocks as sitting on a table. Since blocks touch the table, the table itself provides a ground of Unity.

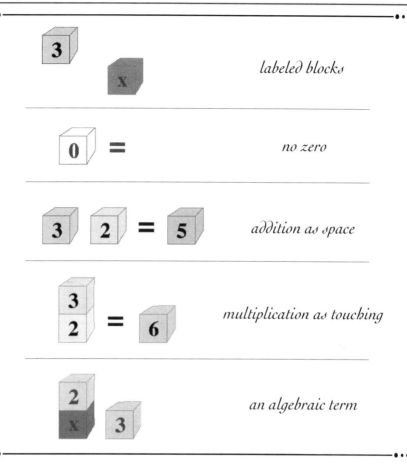

Figure 14-7: *Objects and operations in spatial algebra*

The touching block representation of spatial algebra that models multiplication becomes infeasible when limited to three dimensions, since higher powers (more objects touching) require more spatial dimensions or very convoluted shapes for blocks. The visual metaphor of touching is weak in that it does not convey the interpenetration of the objects implied by multiplication. This problem gets particularly difficult for multiplying several factored expressions, such as (x + 1)(x + 2)(x + 3).

Chapter 14

Distribution

The representation of distribution in spatial algebra is particularly compelling. Blocks with identical labels are both singular and subdividable. Spatial distribution permits combining blocks with identical labels into a single block with the same label. Conversely (read right to left), distribution permits splitting a single block that touches separate piles into separate but identical blocks touching each pile.

distribution

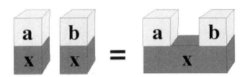

Symbolic string equivalent: ax + bx = (a + b)x

This ability to arbitrarily divide and combine blocks with a common name is the same as the ability to arbitrarily create duplicate labels in a textual representation. Any potential ambiguity between replication due to distribution and replication intended as addition is avoided by the effect of context on interpretation. Distributive replication requires the context of touching blocks (multiplication). Additive replication requires the context of non-touching blocks.

Factoring a polynomial expression is equivalent to multiple applications of distribution. In spatial algebra this means splitting multiple blocks. As an example, here is a factored stack on the right, and the polynomial pieces on the left.

factoring

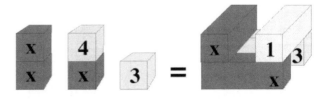

Symbolic string equivalent: $x^2 + 4x + 3 = (x + 1)(x + 3)$

One advantage of the spatial representation on the right is that both the factored and the polynomial forms are visible concurrently. Looking from the side, we see two completely touching spaces that represent the factored form. Looking down at the top, we see four piles that represent the polynomial form. Here, the factored form is converted to the polynomial by cutting each block through the middle.

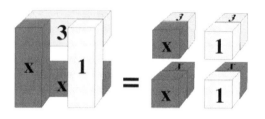

factored to polynomial

Symbolic string equivalent:

$(x + 1)(x + 3) = x^2 + 1x + 3x + 3$

Figure 14-8 on the following page provides an example of using spatial algebra to simplify a deeply nested algebraic expression. The textual notation for algebra requires **grouping parentheses**, but these delimiters are not part of algebra. They are instead a necessary compensation for failure to express the structure of an algebraic expression in sufficient dimensions.[24]

The expression is initially represented as a stack of blocks in *Frame 1*. The first step shown in *Frame 2* is to cut through all the stacked blocks, in effect applying distribution to the entire expression concurrently. Next, as is done in the algebraic simplification, we will assume knowledge of the number facts. Rather than applying them one-by-one, all of the numeric multiplications can occur at the same time in *Frame 3*. Now all that is left is to join like terms. The stacks are rearranged in *Frame 4* due to the artifact of displaying a two-dimensional representation on the page. Again number facts are applied to each different term, all at the same time, shown in

Chapter 14

$$2(x - 3(x - (2y + 1))) - 4(3(y + 1) - x) + 6$$

1.

2.

3.

4.

5.

$$
\begin{aligned}
&2(x - 3(x - (2y + 1))) - 4(3(y + 1) - x) + 6\\
&2(x - 3(x - 2y - 1\ )) - 4(3(y + 1) - x) + 6\\
&2(x - 3x + 6y + 3\ \ ) - 4(3(y + 1) - x) + 6\\
&2(\ \ - 2x + 6y + 3\ \ ) - 4(3(y + 1) - x) + 6\\
&\quad\quad - 4x + 12y + 6\ \ - 4(3(y + 1) - x) + 6\\
&\quad\quad - 4x + 12y\quad\quad - 4(3(y + 1) - x) + 12\\
&\quad\quad - 4x + 12y\quad\quad - 4(\ 3y + 3\ - x) + 12\\
&\quad\quad - 4x + 12y\quad\quad -\ \ 12y - 12 + 4x + 12\\
&\quad\quad - 4x\quad\quad\quad\quad\quad\quad\quad\ \ - 12 + 4x + 12\\
&\quad\quad\quad\quad\quad\quad\quad\quad\quad\quad\quad\quad\ \ - 12\quad\quad + 12\\
&\quad\quad\quad\quad\quad\quad\quad\quad\quad\quad\quad\quad\quad\quad\quad\quad\quad\quad\ \ 0
\end{aligned}
$$

Figure 14-8: *Cutting through symbolic distribution*

Frame 5. In spatial notation, what remains would be erased, leaving nothing at all, i.e. the empty page. The symbolic algebra simplification is also included at the bottom of the figure.

Additional material: iconicmath.com/algebra/spatial/

The **Spatial Algebra** page includes a broader discussion of the material in this section.

14.5 Models of Multiplication

In the final analysis, both multiplication-as-substitution and multiplication-as-structure use substitution to modify structural form. Figure 14-9 shows the rules of each system that are involved with multiplication. In the depth-value approach of multiplication-as-substitution, specific forms are identified and substituted after pattern-matching within a target form (the INTO form). In multiplication-as-structure, a limited number of patterns (the axioms) are identified and substituted after pattern-matching within a target form (the INTO form). If that sounds like the same process, it's because it is.

The difference is that depth-value is essentially an arithmetic technique, it does well with specific numbers. Structural pattern-matching is essentially an algebraic technique, it does well with abstract variables. Both can be used to implement either arithmetic or algebra, and with different boundary shapes, both can be mixed within the same form.

Depth-value narrows the use of substitution by making it necessarily commutative. A non-commuting form cannot be substituted. James algebra allows the use of substitution in its full generality as a match-and-substitute engine for arbitrary pattern axioms.

GROUP

depth-value arithmetic

$$\bullet.._N..\bullet = (\bullet) \quad \Rightarrow \quad 1.._N..1 = N \times 1$$

structural algebra

$$o.._N..o = N \quad \Rightarrow \quad 1.._N..1 = N$$

MERGE

depth-value arithmetic

$$(A)(B) = (A\ B) \quad \Rightarrow \quad (N \times A) + (N \times B) = N \times (A + B)$$

structural algebra

$$A.._N..A = ([A][o.._N..o])$$
$$([A][\quad N\quad]) \quad \Rightarrow \quad A.._N..A = N \times A$$

SUBSTITUTE

depth-value arithmetic

$$(\!(A \bullet E)\!) = (\!(E \bullet A)\!) \quad \Rightarrow \quad A \times E = E \times A$$

structural algebra

$$(\!(([A][M])([A][N]))\ ([A][M\ N])\ E)\!)$$
$$\quad \Rightarrow \quad (A \times M) + (A \times N) = A \times (M + N)$$

Figure 14-9: *Models of multiplication*

Grouping

Both systems accumulate units and give the cardinality of the collection a name (here, N). Depth-value however, introduces an explicit grouping boundary, a **base** that establishes the size of a group. The grouping boundary *defines* the depth-value meaning of multiplication: crossing the boundary outward multiplies the contents by the base. This is not a generic multiplication, it is a specific multiplication by the selected base. In contrast, the structural approach has no independent grouping operation. That's why depth-value can be freely mixed with structural forms. Instead, structural grouping is integrated with structural merging. The structural grouping concept is that of naming the cardinality of a particular collection. There is thus one named structural group for every natural number.

Merging

Depth-value grouping boundaries are merged by the void-based rule:

$$)(= void$$

This simple transformation implements the distributive rule of arithmetic. Group and Merge provide the complete depth-value mechanism for management of magnitude.

In the structural approach Group and Merge are fully integrated within the transformation rule of Replication.

$$A.._N..A = ([A][o.._N..o]) \quad\quad \textbf{replication}$$

The cardinality of a collection of identical structures is determined and then placed within the form of multiplication, together with the replicated structure. This is the *structural definition* of multiplication. The definition is unwound via replication twice, once for the cardinality N and once for each replica of the form A, to yield a unit-ensemble with the cardinality of the product.

Substituting

Both types of multiplication use substitution as an implementation mechanism, however, the concept of multiplication is embodied in very different aspects of each system. It is useful to observe that the concept of generic multiplication is not necessarily dependent upon any one of the three operations of grouping, merging, or arranging. Substitution of unit-ensembles alone is sufficient, as is Replication of forms.

Chapter 14

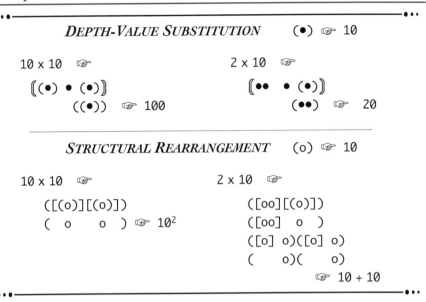

Figure 14-10: *Examples of two models of multiplication*, base-10

Figure 14-10 compares two simple multiplications in base-10 for each of the systems. Both depth-value multiplications are implemented by direct substitution. Both structural multiplications are implemented by substitutions permitted by structural transformation axioms.

14.6 Remarks

Spencer Brown and Kauffman have explored map diagrams extensively using the rule: *Common boundaries cancel*. Kauffman as well has applied boundary techniques quite naturally to knot theory.[25]

Any innovation is the result of the contributions of many. We have explored only the formal aspects of iconic arithmetic, although thinking with distinctions has found explicit advocates in anthropology, sociology, biology, mathematics, computer programming, philosophy and esoterics.

Now it is time to look forward.

Endnotes

1. **opening quote:** L. Kauffman (2011) Laws of Form and topology. *Cybernetics and Human Knowing* **20**(3-4) p.76.

2. **the development of a boundary algebra of numbers in 1961:** Spencer Brown's approach to boundary numbers was distributed as a hand-written manuscript for many years, until finally published as an appendix to a 2009 reprint of *Laws of Form*.

G. Spencer Brown (1969) *Laws of Form*, Bohmeier Verlag Edition 2009 Appendix 4 *An algebra for the natural numbers* (1961).

3. **a given element is or is not contained in another given element:** Spencer Brown, *Laws of Form* p.132.

4. **the dominant error that folks make:** especially apparent in published academic articles and papers.

5. **generalize the arithmetic operations of addition and multiplication:** Wikipedia article *Ring (mathematics)* online June 2016.

6. **cancellation:** Spencer Brown names this rule Reflexion. We are calling it Cancellation here for consistency with Kauffman's terminology. The James algebra name for this rule is Inversion, with the deletion direction of Inversion named `cancellation`.

7. **in his short appendix on numbers:** Spencer Brown, *An algebra for the natural numbers*, Appendix 4.

8. **extended this communality to include finite set theory:** J. Engstrom (2000) *Unifications for Natural Number Arithmetic from a Laws of Form-based Notation* Maharishi University of Management.

9. **that both Lou Kauffman and Jack Engstrom have addressed:** L. Kauffman (1995) Arithmetic in the form. *Cybernetics and Systems: A International Journal* **26** p.1-57.

J. Engstrom (1994) *Natural Numbers and Finite Sets Derived from G. Spencer-Brown's Laws of Form* Master's thesis, Maharishi International University.

Chapter 14

10. **it has no numeric interpretation:** Engstrom (2000) *Unifications for Natural Number Arithmetic from a Laws of Form-based Notation* p.26.

11. **a dual to the Spencer Brown numbers that resolves their interpretative ambiguity:** Kauffman, Arithmetic in the form.

12. **put the contents together in a single container:** Kauffman, p.14.

13. **alternating additive and multiplicative levels:** This perspective is discussed briefly in Chapter 8.

14. **indicate a boundary whose inside is multiplicative:** Kauffman, p.22.

15. **because this expression is not arithmetical:** Kauffman, p.22.

A similar circumstance occurs in James numbers, where [ ] is taken to be non-numeric. In the James system, however, ( )[ ] is accepted as a valid boundary form, one that happens to be interpreted as 1 – ∞.

16. **only if a and b are in opposite universes:** Kauffman, p.22.

17. **our abilities to form patterns and context:** Kauffman, p.21.

18. **inextricably mixed with their own operative powers:** Kauffman, p.30.

19. **depth-value arithmetic described in Chapter 3:** Kauffman's earlier work on boundary arithmetic includes:

L. Kauffman (1985) *Sign in Space*. First Annual Conference on Sign and Space, Santa Cruz

L. Kauffman (1986) *Formal Arithmetic*. Department of Mathematics, Statistics and Computer Science, University of Illinois at Chicago.

L. Kauffman (1987) *The Form of Arithmetic*. 18th International Conference on Multivalued Logic.

I recommend his boundary math work assembled in *Laws of Form — An Exploration in Mathematics and Foundations* (rough draft). Online 2/17 at http://homepages.math.uic.edu/~kauffman/Laws.pdf

20. **when place-marker boundaries collapse through the rule)(= void:** Kauffman, Arithmetic in the form p.38.

21. **rules handle multiplication of negative numbers:** Kauffman, *The Form of Arithmetic*.

22. **a learning tool for those first encountering algebra:** The difficulties students have when they begin to learn algebra are well documented. Spatial algebra addresses common errors made by novice algebra students by permitting experiential interaction with virtual forms. Spatial representations enhance understanding. Concrete manipulation is known to be an effective teaching technique for novices.

The educational approach is to anchor concepts first to experience, and then later to introduce the more abstract symbolic concepts. For novice algebra students, the symbolic form is more prone to errors since it ignores learning theory. We can say that the structure of symbolic algebra *affords*, or facilitates, student errors. Conceptual understanding is better supported by iconic systems that can be physically manipulated.

The iPad/iPhone application Wolfram Algebra Course Assistant from Wolfram Alpha uses Mathematica to perform all algebraic manipulations, freeing the student to focus on using algebra rather than doing computation. From a modern digital perspective, we can now leave all computing to the computers that we built for precisely that purpose. Having students learn algebraic manipulations in the twenty-first century is analogous to asking them to ride a horse to school rather than taking the school bus.

The question of empirical evaluation of spatial algebra for improving math education is subtle. Spatial algebra teaches a different set of concepts than does symbolic group theory. If success in school means understanding the manipulation of symbols based on abstract rules, then spatial algebra may even reduce competence. The problem, however, is with the criteria for success, not with student learning. Schools have a great difficulty updating concepts and approaches that have been obsoleted by advancing culture and technology. Manipulation of symbols itself is an anachronism, so that establishing this skill as the criterion for success in high school algebra is plainly dysfunctional. Many modern improvements in math education speak of conceptual learning rather than memorization of algorithms. However, today still over 80% of math classrooms are mired in what is rapidly becoming ancient history. Symbolic algebra lacking computer support for computation is as dead as learning ancient Greek and as intellectually misguided as studying the phlogiston theory of combustion.

Chapter 14

23. **Numbers and variables are represented by physical blocks:** This content was first published in W. Bricken (1992) Spatial representation of elementary algebra, *1992 IEEE Workshop on Visual Languages* p.56-62. Spatial algebra was also presented at the 1992 meeting of the American Educational Research Association in the context of designing a virtual reality math learning environment for algebra students.

Daniel Shapiro spent a summer implementing an immersive virtual environment for doing boundary logic as part of this project.

W. Winn & W. Bricken, Designing virtual worlds for use in mathematics education: The example of experiential algebra. *Educational Technology*, **32**(12), p.12-19.

24. **failure to express the structure of an algebraic expression in sufficient dimensions:** Every arithmetic student must learn the notorious PEMDAS rules for disambiguating the sequence of arithmetic operations. An expression such as 4 x 3 + 2 is ambiguous without grouping parentheses. The intent of the expression could be either 4 x (3 + 2) or (4 x 3) + 2. Students must memorize an order of operations to compensate for the laziness of whomever wrote down the expression. When expressed in spatial algebra, or in a two-dimensional graph, there is no ambiguity. PEMDAS is not mathematics, it is a notational patch that covers a poor conceptualization of both arithmetic and communication.

25. **applied boundary techniques quite naturally to knot theory:** The idea that common boundaries cancel was named **idemposition** by Spencer Brown and Kauffman. When expressed in a delimiter notation such as parentheses, the rule invites us to transcend the left/right delimiter structure.

$$)(= \textit{void}$$

See L. Kauffman (2005) Reformulating the map theorem, *Discrete Mathematics* 302 p.145-172. Online 1/17 at http://arxiv.org/pdf/math/0112266.pdf and on Kauffman's homepage, http://homepages.math.uic.edu/~kauffman/Papers.html. For a less technical description, see L. Kauffman (2011) Laws of Form and topology. *Cybernetics and Human Knowing* **20**(3-4) p.50-100.

Kauffman's extensive use of boundary math techniques in knot theory can also be found on his webpage and compiled in L. Kauffman (1993) *Knots and Physics*.

Chapter 15

Next

In the brain there is no principled distinction between hardware and software or, more precisely, between symbols and nonsymbols.[1]
— *Francisco Varela (1992)*

Our exploration has been conducted on an unlevel playing field. We have been using symbols to raise awareness of the postsymbolic nature of thought, and we have undertaken the exploration within a most unfriendly territory, that of symbolic arithmetic. An implicit expectation has been that the reader is willing to explore iconic form while also considering that the thoughts engendered might be free of symbolic reference. Thought without words, experience without chatter? Distinction, that stuff of minds, requires difference, not reference.

15.1 Choice

How we think about mathematical concepts is influenced by how those concepts are presented and represented. Syntax and semantics, representation and meaning, are tightly connected. In general, how we record and manipulate numbers is a matter of convenience, but the convenience of the *learner* may have been forgotten. For learning mathematics — and for using mathematics — it

Chapter 15

is more convenient to call upon sensory interaction and natural behavior than it is to manipulate symbols.

A purpose of this volume is to provide evidence that our cultural and academic commitment to symbol processing is a *design choice* and not an inevitability. We have explored two different formal models of arithmetic that are iconic rather than symbolic. Well, we have actually brushed the surface. Under each iconic structure there lies a deep well of potential innovation and opportunity waiting for an intrepid explorer whose appetite may have been whetted by the suggestion that numbers are far more than numerals and symbolic transformation rules. Numbers are contextual relations.

A premiere design concern is comprehension by non-professionals, particularly students. If we did not have to conform to prior instruction, what would be the most desirable way to help students learn how arithmetic works? Just as important as student learning is an overarching question. To what extent have the recent technological and electronics revolutions changed our understanding from a century ago of what arithmetic is? For a society that inundates itself with high density visual information at every waking moment,

It is no longer reasonable to claim that cognitive skill lies in typographic symbols.

Underlying our mutual exploration is the overt observation that arithmetic is far broader, conceptually and experientially, than what is taught in schools and, indeed, what is thought throughout an academic culture focused on symbolic markings. If indeed math is important to learn, for rigor and for clarity, then surely we must question *which type of math* is important for organic beings. And which dialects are appropriate for the 3D digital age? Does *symbolic rigor* mean preparation for the future, or is it perhaps a history lesson?

Human Nature

A more radical suggestion is that our twentieth century excursion into symbolic mathematics has been a temporary transition at best and a dimensionally degenerate delusion at worst. Each of us is born with evolutionary perspective and with organic knowledge of what might be called **humane math**. Neuroscientist Dehaene is specifically critical of formalist definitions.

> Ironically, any 5-year-old has an intimate understanding of those very numbers that the brightest logicians struggle to define. No need for a formal definition. We know intuitively what integers are. Among the infinite number of models that satisfy Peano's axioms, we can immediately distinguish genuine integers from other meaningless and artificial fantasies. Hence our brain does not rely on axioms.[2]

Mathematics does not necessarily describe nature, it describes our human nature to occlude, to abstract and to simplify. Math is the tool that our culture uses to keep Reality from being overwhelming. Mathematician and historian Morris Kline:

> Mathematics is not something independent of and applied to phenomena taking place in an external world but rather an element in our way of conceiving the phenomena. The natural world is not objectively given to us. It is man's interpretation or construction based on his sensations, and mathematics is a major instrument for organizing the sensations.[3]

There are many thoughts, many sensory modes and many mathematics. The question is not which is right, or even which is better. The question is how do we wish to view ourselves?

Chapter 15

15.2 A Hidden Motive

There has been, all along, a hidden agenda, originally initiated by Spencer Brown. The idea is to construct the foundations of logic, numerics and sets, basically all of finite mathematics, from the *same* boundary concepts and forms. Spencer Brown's *Laws of Form* for iconic logic are

crossing

⟨⟨ ⟩⟩ =

Crossing

The value of a crossing made again is not the value of the crossing.

calling

⟨ ⟩⟨ ⟩ = ⟨ ⟩

Calling

The value of a call made again is the value of the call.

Thomas McFarlane observes fairly that James algebra lacks a clear connection to Spencer Brown's *Laws of Form*.

accumulating

( )( ) ≠ ( )

> The transformative rules and various types of boundaries are introduced *ad hoc* without providing any intuitive basis for them. What is the basis for the adoption of three different boundaries? Are there more fundamental justifications for the axioms governing the transformation of expressions? Is there a deeper connection with Spencer-Brown's arithmetic?[4]

McFarlane brings to attention that there are both mathematical and philosophical motivations in the study of distinction and in *Laws of Form*. In *Distinction and the Foundation of Arithmetic* he derives the James axioms from intuitive first principles. McFarlane's argument is philosophical whereas the presentation herein is structural. It is this structural mapping that awaits a deeper study.

Arithmetic and Logic

Each version of arithmetic in this volume has remained as close as possible to Spencer Brown's formulation of logic, pivoting on a single modification. In *Laws of Form* Spencer Brown constructs elementary logic from two

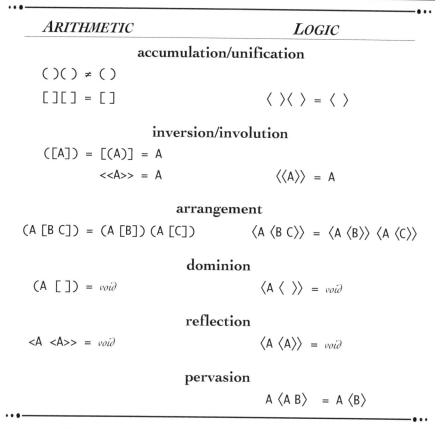

Figure 15-1: *Boundary arithmetic and boundary logic*

axioms. The **Law of Calling** generates logical form by suppressing accumulation, while in the arithmetic of numbers Accumulation, the denial of Calling, generates numeric form. Calling allows repetition that does not change value, while Accumulation identifies a specific type of repetition, the repetition of replica units, that in effect *defines value*. The **Law of Crossing** holds for both arithmetic and logic. Crossing defines how value is changed rather than how it is created. Value changes when we cross a boundary.

Figure 15-1 provides a brief comparison. The **logical boundary** is rendered in the figure as ⟨ ⟩. The small differences in the structure of boundary arithmetic and boundary logic provide substantive clues about the foundational structure of finite mathematics. Since boundaries represent cognitive distinctions, the difference in the structure of logic and numeric form specifically identifies the difference between thinking logically and thinking numerically.

Construction

Arithmetic is constructed from logic by these modifications.

Units exist.

$$(\ ) \neq \qquad [\ ] \neq \qquad \langle\ \rangle \neq$$

Arithmetic consists of **two types of units**. One accumulates and one is the same as a logic unit. Logical Truth is a type of numeric infinity.

$$(\ )(\ ) \neq (\ ) \qquad [\ ][\ ] = [\ ] \qquad \langle\ \rangle\langle\ \rangle = \langle\ \rangle$$

Inversion prohibits mutual containment of the two types of numeric unit, as it does for the logic unit.

$$([\ ]) = [(\ )] = \qquad \langle\langle\ \rangle\rangle =$$

Nested pairs do not condense, just like the logic unit.

$$((\ )) \neq (\ ) \qquad [[\ ]]\ \textit{undefined} \qquad \langle\langle\ \rangle\rangle \neq \langle\ \rangle$$

Arrangement is the same for both numeric inversion frames and the logic boundary.

$$(A\ [B\ C]) = (A\ [B])\ (A\ [C])$$
$$\langle A\ \langle B\ C\rangle\rangle = \langle A\ \langle B\rangle\rangle\ \langle A\ \langle C\rangle\rangle$$

Dominion defines a hierarchy of existence.

$$(\ )[\ ] = [\ ]$$

All of this mechanism is designed to maintain accumulation of arithmetic units only, and to assure otherwise that arithmetic units do not get replicated during transformation and rearrangement. Arithmetic units cannot cross a boundary without changing the cardinality of the boundary. Within this constraint arithmetic units are indistinguishable from logic units. When the angle bracket is introduced, it too is equivalent to the logic boundary.

The other substantive difference between arithmetic and logic is that the logic boundary is semipermeable. **Pervasion** asserts that replicas of outside forms can cross a logic boundary to the inside, something that is forbidden for any type of numeric boundary. Semipermeability gives limited permission to replicate without changing cardinality. Not surprising since logic does not support accumulation of cardinality.[5]

pervasion

A ⟨A B⟩ = A ⟨B⟩

15.3 It's Not Easy

Reflection upon the history of the development of mathematical concepts, as John Derbyshire writes in his history of algebra, will "make us realize how deeply unnatural mathematical thinking is."[6] Not only do we have no evolutionary adaptation or propensity for purely abstract thinking, also the greatest intellects of the last two millennia have struggled mightily with what we believe today should be common mathematical knowledge. Here's Morris Kline again:

> In retrospect, this glorification of mathematical reasoning seems incredible. To be sure, tatters of reasoning were employed. But especially in the 18th century when heated debates about the meaning and properties of complex numbers, logarithms of negative and complex numbers, the foundations of the calculus, the summation of series, and other issues we have not described filled the literature, the designation Age of Confusion seems more appropriate.[7]

Symbolic arithmetic is not easy, it has taken centuries for humanity to develop it. John Derbyshire:

> The extreme slowness of progress in putting together a symbolic algebra testifies to the very high level at which this subject dwells. The wonder ... is not that it took us so long to learn how to do this stuff; the wonder is that we can do it at all.[8]

Chapter 15

Math Education

The point of demarcation is that students in grade school and in high school and in college do not need to understand arithmetic from the perspective of a foundational mathematician, of which there are only a few hundred world-wide. That would be like insisting that Xtreme athletes import their training regimes into grade school playgrounds, and that children in science classes emulate the life-long research strategies of Nobel laureates. What is misunderstood by mathematics educators is that expertise requires both a vast array of baseline knowledge and extensive training in applying that knowledge. Children, and adults too, are not mathematical experts but the notations, computational strategies and modes of thought incorporated into high school math are intended to emulate those of mathematical experts.

The problem is that professional mathematical tools have leaked down into elementary math education as if they were arithmetic itself. Prior to anchoring our concept of number to sets and to logical theory about a century ago, there was another much simpler arithmetic based on the intuitive Additive Principle. It is this organic understanding of numbers that should be taught in schools. Logic and sets do not have an exclusive right to claim to be the *only* formal foundation for numeric arithmetic.

There is yet another perspective: school math classes do not teach *math*, or at most teach only the tiniest portion of actual mathematics. Mathematician Ian Stewart:

> What mathematics is, and how useful it is, are widely misunderstood. It is not solely about numbers, 'doing sums' as we were taught in school — that's arithmetic. Even when you add in algebra, trigonometry, geometry and various more modern topics such as matrices, what we learn in school is a tiny, limited part of a vast enterprise. To call it one-tenth of one per cent would be generous.[9]

What is taught is not math but rather computer science, the domain that studies and characterizes the behavior of algorithms and automatons. The rigidly structured curriculum materials and the standardized tests of algorithmic skill and technocultural fact is what computer science calls software programming. Math education attempts to download a silicon programming language into organic beings while completely ignoring teaching the skills of symbolic programming.

For an understanding of mathematics, for any hope that future students will be receptive to mathematics, for the possibility of teaching mathematics well, indeed for the possibility of teaching mathematics at all, we need to return to the simple conceptual foundation of mathematics as direct physical experience.

15.4 Summary

Figure 15-2 collects the explicit principles of boundary arithmetic that are scattered throughout the chapters. These principles provide structure for a conceptual map of learning and teaching elementary arithmetic. The principles are stated succinctly so that they may stand alone as bumper-stickers. Some of the phraseology is new, in order to connect the concepts to the central organizing idea of *making a distinction*.

What can be said in summary? Sums are fusions. What is the fusion of the content of this volume? Fusions are wholes. We have drawn a boundary, made a single distinction. Outside there is experience; inside, thought.

You can tell if you understand a mathematics by changes in your vision.

Inside, where there is nothing, is measured by change on the outside. Outside is identical to the distinctions that we elect to construct.

Chapter 15

	VOID	*page*
Void	Void has no properties.	15
Existence	Something is not nothing.	168
	MEANING	
Distinction	Difference is an idea.	15
Calling	Repetition does not make a difference.	372
Crossing	Crossing a boundary makes a difference.	372
Axiomatic	If it is not explicitly allowed, it is forbidden.	140
Communality	When it is shared by all content, it is context.	36
Semantics	A pointer is not what is pointed at.	213
Participation	Meaning depends on how we look.	13
Void-equivalent	Void-equivalent structure cannot make a difference.	151
	ARITHMETIC	
Accumulation	Parts accumulate rather than condense.	171
Hume's	Equality is one-to-one correspondence.	62
Additive	A sum looks like its parts.	3
Multiplicative	Each part of one touches each part of the other.	3
Arrangement	Arrangement is the sole source of complexity.	197

Figure 15-2: *The principles of boundary arithmetic*

From Here

There are two volumes on iconic arithmetic to follow. In Volume II we address just what formalism is, as a mathematical philosophy and as a computational paradigm. We'll go back to visit the original models of arithmetic developed by Frege, Peano, Robinson and others to compare the formal structure of symbolic expressions and iconic form. We have glossed lightly over the concept of **equality**, so we will mold it into void-equivalence. Similarly we have yet to integrate parallelism into our

Next

formal model. Then it will be time to address *symbolic and iconic metamathematics*.

We'll explore pure boundary arithmetic, its internal structure and formal consequences without regard to interpretation. Just how can arithmetic fit neatly into one binary relation? What are the structural properties of containment? Just what have we taken on in claiming that arithmetic is about a single physical relationship?

We will be able to dismiss the angle-bracket entirely, reducing James algebra to two boundary types without loosing the integration of operations and their inverses. This of course raises further questions about group theory, since inverses too will be removed from the current theory of arithmetic, reduced to a notational abbreviation. Only Distribution remains, to allow smooth transition between addition and multiplication (and incidentally to permit the generation of mathematical complexity).

Volume II then feels quite different, with deeper, more challenging questions at the foundations of the current philosophies of mathematics. This is a necessary volume to address the many technical details about the structures, assumptions and thought processes that we now expect grade school teachers and students to grasp intuitively.

In Volume III we'll tackle the neglected topic of the empty square-bracket and its possible interpretation as an infinity. What are the consequences of mixing numeric and non-numeric units? How does James algebra handle division-by-zero and indeterminate forms and exotic bases? These noxious concepts are in the interpretation, but what are they in the form? We will be able to organize the indeterminate forms such as $\infty - \infty$ and $\infty/0$ into a single coherent pattern. The Mother of all imaginaries, -1, is implicated with every strangeness that occurs in arithmetic.

Volume III returns to pure exploration of form by examining J, the logarithm of −1, as the foundation of imaginary numbers from which our current compound imaginary i can be derived. Just what does the imaginary numeric realm mean, and what are its fundamental structures? We'll take a close historical look at Euler and Leibniz and Bernoulli as they invent complex numbers yet fail to converge on the meaning of J. We will see the features of the imaginary domain arise out of a simpler foundation guided by patterns of containment.

We'll revisit the oscillation of logarithmic and exponential levels of nesting in the context of imaginary forms, and take a closer look at the concept of mathematical morphism. And then we'll unify π, trigonometry, cyclic behavior, hyperbolic functions and complex logarithms naturally within iconic form.

Phew! And hopefully a solid foundation for beginning to understand, in Volumes IV and V, Spencer Brown's reconceptualization of *logical thought*. Boundary logic itself provides a far more revolutionary reconstruction of the nature of rationality than does our current exploration of boundary arithmetic. Logic is unary not dualistic; FALSE is a void-equivalent concept that can be completely disregarded and discarded. Deduction means to identify and delete void-equivalent forms. The path to critical thinking (as well as to new computational architectures) is through removal rather than accumulation of structure. The basis of rationality is emptiness.

15.5 Remarks

Spencer Brown's book *Laws of Form* is seminal, but in the fifty years since it was written our experience with boundary logic has grown significantly. There is an extensive collection of papers describing boundary logic at www.iconicmath/logic/boundary/.

This volume is if anything pragmatic. And yet some philosophy shows through in the form of the metaphysics of *void*. There are two voids. One we can talk about; that one is indicated by its boundary. That void is coupled to distinction as the foundation of unity. The other void is destroyed upon mention. The unmentionable is the metaphysical motivation of this volume. The concept of number is already so far removed from its origin that it is essential to regress backwards, from Two to One to Nothing to Silence, in order to find number, unity and absence.

Endnotes

1. **opening quote:** F. Varela (1992) *Ethical Know-How* p.54.

2. **the brain does not rely in axioms:** S. Dehaene (2011) *The Number Sense: How the mind creates mathematics* p.223.

3. **mathematics is a major instrument for organizing the sensations:** M. Kline (1980) *Mathematics The Loss of Certainty* p.341.

4. **Is there a deeper connection with Spencer-Brown's arithmetic?:** T. McFarlane (2007) *Distinction and the Foundations of Arithmetic.* Online 6/16 at http://www.integralscience.org/tom/

5. **logic does not support accumulation of cardinality:** For an extremely concise summary of the growth of arithmetic and logic from nothing at all, see W. Bricken (2006) The Mathematics of Boundaries: A beginning. In D. Barker-Plummer *et al* (eds.) *Diagrams 2006*, LNAI 4045 p.70-72.

6. **how deeply unnatural mathematical thinking is:** J. Derbyshire (2006) *Unknown Quantity* p.40.

7. **the designation Age of Confusion seems more appropriate:** Kline, p.169.

8. **the wonder is that we can do it at all:** Derbyshire, p.51.

9. **to call it one-tenth of one per cent would be generous:** I. Stewart (2011) *Mathematics of Life* p.8.

Chapter 15

Bibliography

All entries in the bibliography are from the chapter endnotes.

W. Allen (1988) in H. Eves *Return to Mathematical Circles*

Aristotle (2015) *Protrepticus*, reconstructed by D. Hutchenson & M. Johnson

V. Arnold (1998) On teaching mathematics. A. Goryunov (trans.) *Russian Math. Surveys* **53**(1) p.229-236

Z. Artstein (2014) *Mathematics and the Real World*

R. Augros & G. Stanciu (1984) *The New Story of Science*

A. Badiou (2008) *Number and Numbers*

J. Barrow (2000) *The Book of Nothing*

J. Barwise & J. Etchemendy (1996) Visual information and valid reasoning. In G. Allwein & J. Barwise (eds.) *Logical Reasoning with Diagrams*

G. Bateson (1972) *Steps to an Ecology of Mind*

_____ (1991) *A Sacred Unity*

G. Boolos (1998) *Logic, Logic, and Logic*

N. Bourbaki (1950) The architecture of mathematics. *American Mathematical Monthly* v57

W. Bricken (1987) *Analyzing Errors in Elementary Mathematics*. Doctoral dissertation. Stanford University School of Education.

_____ (2006) The mathematics of boundaries: A beginning. In D. Barker-Plummer *et al* (eds.) *Diagrams 2006*, LNAI 4045, p.70-72

L. Bunt, P. Jones & J. Bedient (1976) *The Historical Roots of Elementary Mathematics*

M. Burns (1998) *Math: Facing an American phobia*

F. Cajori (1928) *A History of Mathematical Notations*

R. Calinger (1999) *A Contextual History of Mathematics*

R. Carnap (1937) *The Logical Syntax of Language*

L. Carroll (1871) *Through the Looking Glass*

Bibliography

J. Conway (1976) *On Numbers and Games*

R. Courant & H. Robbins (1969) *What is Mathematics?*

U. D'Ambrosio (2000) A histographical proposal for non-western mathematics. In H. Selin (ed.) *The History of Non-western Mathematics* p.131-158

P. Davis (1997) Mathematics in an age of illiteracy. *SIAM News* **30**(9) 11/9

_____ (2004) A brief look at mathematics and theology. *The Humanistic Mathematics Network Journal Online* 7

P. Davis & R. Hersh (1981) *The Mathematical Experience*

DaVinci (1954) *The Notebook* translated and edited by E. Macurdy

R. Dedekind (1872) *Continuity and Irrational Numbers*

J. Derbyshire (2006) *Unknown Quantity*

S. Dehaene (2011) *The Number Sense: How the mind creates mathematics*

K. Devlin (2000) *The Math Gene*

_____ (2006) The useful and reliable illusion of reality in mathematics. *Toward a New Epistemology of Mathematics Workshop*, GAP6 Conference

_____ (2011) *Mathematics Education for a New Era*

P. Dirac (1963) The evolution of the physicist's picture of nature. *Scientific American* **208**(5) p.45-53

A. Einstein (1977) *Reader's Digest* 10/1977

_____ in P. Schilpp (ed.) (1979) *Autobiographical Notes. A Centennial Edition* In D. Howard & J. Stachel (2000) *Einstein: The Formative Years, 1879-1909*

J. Engstrom (1994) *Natural Numbers and Finite Sets Derived from G. Spencer-Brown's Laws of Form* Master's thesis, Maharishi International University

_____ (2000) *Unifications for Natural Number Arithmetic from a Laws of Form-based Notation* Maharishi University of Management

Euclid (c. 300 BCE) *The Elements*

L. Euler (1802) Letter CIV Different forms of syllogisms, (2/21/1761), H. Hunter (trans.) *Letters of Euler*

L. Euler et al (1822) *Elements of Algebra*

R. Feynman (1985) in K. Cole *Sympathetic Vibrations: Reflections on physics as a way of life*

Bibliography

G. Frege (1884) *The Concept of Number*

_____ (2004) in M. Potter *Set Theory and its Philosophy*

M. Gell-Mann (1980) in H. Judson *Search for Solutions*

H. Genz (1999) *Nothingness: The science of empty space*

J. Goguen (1993) *On Notation*. Department of Computer Science and Engineering, University of California at San Diego

T. Gowers (2005) Does Mathematics Need a Philosophy? Online 6/16 at https://www.dpmms.cam.ac.uk/~wtg10/philosophy.html

E. Gray & D. Tall (1994) Duality, ambiguity and flexibility: A proceptual view of simple arithmetic. *Journal for Research in Mathematics Education* **26**(2) p.115-141

A. Grothendieck (2011) in R. Hersh and V. John-Steiner *Loving + Hating Mathematics*

M. Hallett (1994) Hilbert's Axiomatic Method and the Laws of Thought. In A. George (ed.) *Mathematics and Mind*

Y. Harari (2015) *Sapiens: A brief history of humankind*

G. Hardy (1941) *A Mathematician's Apology*

S. Hawking (2007) *God Created the Integers: The mathematical breakthroughs that changed history*

S. Hawking & R. Penrose (1996) *The Nature of Space and Time*

R. Heck (2000) Cardinality, counting, and equinumerosity. *Notre Dame Journal of Formal Logic* **41**(3)

H. vonHelmholtz (1887) *Counting and Measuring*

V. Huber-Dyson (1998) On the nature of mathematical concepts: Why and how do mathematicians jump to conclusions? EDGE conversation 2/15/98. Online 8/16 at https://www.edge.org/conversation/verena_huber_dys-on-on-the-nature-of-mathematical-concepts-why-and-how-do-mathematicians

G. Ifrah (2000) *The Universal History of Numbers*

M. Johnson (1987) The body in the mind. In F. Varela, E. Thompson & E. Rosch (1991) *The Embodied Mind*

L. Kauffman (1985) *Sign in Space* First Annual Conference on Sign and Space, Santa Cruz

Bibliography

_____ (1986) Formal Arithmetic. Department of Mathematics, Statistics and Computer Science, University of Illinois at Chicago

_____ (1987) The form of arithmetic. *18th International Conference on Multivalued Logic*

_____ (1993) *Knots and Physics 2nd ed.*

_____ (1995) Arithmetic in the form. *Cybernetics and Systems: A International Journal* **26** p.1-57

_____ (2011) Laws of form and topology. *Cybernetics and Human Knowing* **20**(3-4)

_____ (2016) What is a number? Online 6/16 at http://homepages.math.uic.edu/~kauffman/NUM.html

_____ (2017 in process) *Laws of Form — An Exploration in Mathematics and Foundations* (rough draft) Online 2/17 at http://homepages.math.uic.edu/~kauffman/Laws.pdf

M. Kline (1980) *Mathematics The Loss of Certainty*

A. Korzybski (1933) *Science and Sanity: An introduction to non-Aristotelian systems and general semantics*

I. Lakatos (1976) *Proofs and Refutations: The logic of mathematical discovery*

G. Lakoff & M. Johnsen (2003) *Metaphors We Live By*

G. Lakoff & R. Núñez (2000) *Where Mathematics Comes From: How the embodied mind brings mathematics into being*

J. Lanier (2010) *You are Not a Gadget*

J. Larkin & H. Simon (1987) Why a diagram is (sometimes) worth ten thousand words. In B. Chandrasekaran *et al* (1995) *Diagrammatic Reasoning*

Leibniz to Tschirnhaus, (1679) Math., IV, 481; Brief., I, 405. In L. Couturant *The Logic of Leibniz*

M. Leng (2010) *Mathematics and Reality*

J. Littlewood (1986) in B. Bollobás (ed.) *Littlewood's Miscellany*

P. Lockhart (2012) *Measurement*

D. Macbeth (2009) Meaning, use, and diagrams. *Ethics and Politics* **xi**(1) p.369-384

E. Mach (1895) *Popular Science Lectures* T. McCormack (trans.)

A. Martínez (2006) *Negative Math*

_____ (2012) *The Cult of Pythagoras*

H. Maturana & F. Varela (1987) *The Tree of Knowledge: The biological roots of human understanding*

B. Mazur (2003) *Imagining Numbers*

J. Mazur (2014) *Enlightening Symbols*

T. McFarlane (2007) *Distinction and the Foundations of Arithmetic*. Online 6/16 at http://www.integralscience.org/tom/

O. Neugebauer (1962) *The Exact Sciences in Antiquity*

C. S. Peirce (1872) MS. 179, in V. Garnica *Changes and Chances: an initial study of Peirce's pragmatism and mathematical writings as they relate to education and the teaching and learning of mathematics*

_____ (1931-58) *Collected Papers of Charles Sanders Peirce*. Hartshorne, Weiss &Burks (eds.)

_____ (1976) The new elements of mathematics. In C. Eisele (ed.) *Mathematical Philosophy*

R. Penrose (2004) *The Road to Reality*

Plato (c. 400 BCE) *Phaedo*

M. Potter (2004) *Set Theory and its Philosophy*

B. Rotman (1987) *Signifying Nothing: The semiotics of zero*

_____ (1993) *Ad Infinitum The Ghost in Turing's Machine: Taking God out of mathematics and putting the body back in*.

_____ (2000) *Mathematics as Sign: writing imagining counting*

R. Rucker (1987) *Mind Tools*

B. Russell (1923) Vagueness. *Australasian Journal of Philosophy and Psychology* (1) p.84-92

_____ (1956) in J. Newman (ed.) *The World of Mathematics*

J.P. Sartre (1938) *Nausea*. L. Alexander (trans.)

V. Sazonov (1995) On feasible numbers. In D. Leivant (ed.) *Logic and Computational Complexity* LNCS 960

D. Schmandt-Besserat (1987) Oneness, twoness, threeness. *The Sciences* 27 p.44-48

_____ (1992) *How Writing Came About*

Bibliography

H. Simon (1995) in B. Chandrasekaran *et al*, (eds.) *Diagrammatic Reasoning*

Brian Smith (1996) *On the Origin of Objects*

G. Spencer Brown (1969) *Laws of Form*

_____ (1969) *Laws of Form*, Bohmeier Verlag Edition 2009 Appendix 4 *An algebra for the natural numbers* (1961)

Irving Stein (1996) *The Concept of Object as the Foundation of Physics*

I. Stewart (1995) *Nature's Numbers*

_____ (2011) *Mathematics of Life*

F. Varela (1992) *Ethical Know-How*

R. Vithal & O. Skovsmose (1997) The end of innocence: a critique of 'ethnomathematics'. *Educational Studies in Mathematics* **34**

H. Weyl (1959) Mathematics and the laws of nature. *The Armchair Science Reader*

J. Wheeler (1988) World as system self-synthesized by quantum networking. *IBM Journal of Research and Development* **32**(1)

A. Whitehead (1958) *An Introduction to Mathematics*

A. Whitehead & B. Russell (1910) *Principia Mathematica*

A. Wilden (1972) *System and Structure: Essays in communication and exchange*

L. Wittgenstein (1930) *Philosophical Remarks*

_____ (1933) *Philosophical Grammar*

S. Wolfram (2002) *A New Kind of Science*

_____ (2007) *Mathematica Notation: Past and Future*. Online 8/16 at http://www.stephenwolfram.com/publications/mathematical-notation-past-future/

Index to the Index

The index is organized by concept and by content. Keywords are associated with each index category.

As a suggestion, decide first what kind of things you are looking for: *person, concept, pattern*, or *iconic system*. Then look under the appropriate category for page and figure locations. Figures are listed by page number followed by the figure number in bold italics, for example: 124 *5-2*. Boldface page numbers indicate primary definitions.

PEOPLE

SYMBOLIC CONCEPTS
arithmetic
mathematics
number
representation

ICONIC CONCEPTS
boundary thinking
containment
iconic form
maps
principles

ENSEMBLES
block arithmetic
circle arithmetic
depth-value
ensemble arithmetic
iconic calculator
network arithmetic
parens arithmetic
pattern equations

JAMES ALGEBRA
dialects
James form
pattern equations

OTHER ICONIC SYSTEMS
boundary logic
Kauffman arithmetics
spatial algebra
Spencer Brown numbers
volume II
volume III
symbols and icons

COVER
cover words

WEBSITES

Index

PEOPLE
quotations in bold

Woody Allen	**282**
Aristotle	**163**, 164, 179
Vladimir Arnold	159
Zvi Artstein	**18**, 30
Michael Atiyah	**235**
R. Augros & G. Stanciu	**19**, 30
Augustine	164
Alain Badiou	**31**, **176-177**, 181
John Barrow	179
J. Barwise & J. Etchemendy	**9**, 27, **136**, 157
Gregory Bateson	**14**, **15**, 29-30, 238
Harry Belafonte	**33**, 61
Bernoulli	380
Bolyai	18
The Borg	**259**
Nicholas Bourbaki	**8**, 27
William Bricken	30, 367-368, 381
Brouwer	210
L. Bunt, P. Jones & J. Bedient	**158**
Marilyn Burns	83
Florian Cajori	84, 304
Ronald Calinger	282
Rudolf Carnap	**7**, 27
Lewis Carroll	61
John Horton Conway	154, 160
R. Courant & H. Robbins	**163**, 178
Ubiratan D'Ambrosio	28
DaVinci	**179**
Philip Davis	6, 7, 27
P. Davis & R. Hersh	210, **215**, 235, 237
Dedekind	63, 177, 209, **273**-274, 276, 284
Stanislas Dehaene	21, **31**, **279**, **280**-281, 285, **371**, 381
Democritus	163
John Derbyshire	**375**, 381
Descartes	6
Keith Devlin	**84**, **166**, 180, **296**, 304, **307**, 340
Paul Dirac	**19**, 30
Einstein	**18**, 30, **119**, 125
Jack Engstrom	206, 346-347, 365-366
Euclid	135, 157
Euler	**64**, **134**-135, 157, 380
Richard Feynman	**1**, 27
Fibonacci	296
Frege	4, 10, 39, **42**, 58, 62-63, 135, 177, **183**, 204, 209, 378
Galileo	**164**
Vincente Garnica	62
Gauss	18
Murray Gell-Mann	**19**, 30
Henning Genz	**163**, 178
Joseph Goguen	**9**, 28
Timothy Gowers	115, 118
E. Gray & D. Tall	158
Alexander Grothendieck	**155**, 160
Yuval Harari	179
Godfrey Hardy	**19**, 30
Stephen Hawking	235
S. Hawking & R. Penrose	**20**, 30
Richard Heck	**208**, **212**, 235, 237
Helmholtz	215, 237
R. Hersh & V. John-Steiner	160
Hilbert	**7-8**, 27, 63, 135
Verina Huber-Dyson	276, 284
Hume	62, 122
Georges Ifrah	**33**, 61, **65**, 83

Index

Jeffrey James	133, 259		**159**, 212, 219, 233, 342
Mark Johnson	30, 179	Roger Penrose	**19**, 30
Louis Kauffman	**38**, 53, 62-63, 67, 85, 133, 313, **340**, **341**-342, 345-346, **348**-355, **350**, **351**, **352**, **354**, 364-366, 368	Plato	158, **181**
		Michael Potter	63, 235
		Pythagoras	304
		Rafael Robinson	4, 378
Morris Kline	**371**, **375**, 381	Brian Rotman	**23**, 28, 31, **33**, 61, **164**, 179, **207**, **218**, 235, 238, **275**, 280, 284, **305**, 340
Alfred Korzybski	237		
Kronecker	206, 210, **235**		
Lagrange	7		
Imre Lakatos	**17**, 30	Rudy Rucker	**169**, 181
G. Lakoff & R. Núñez	**12**, **13**, **16**, 28-30, 267-**268**, 282	Bertrand Russell	**62**, **136**, 157, **200**
		Jean-Paul Sartre	**161**, 178
Jaron Lanier	5, **27**	Vladimir Sazonov	285
J. Larkin & H. Simon	**135**, 157	Denise Schmandt-Besserat	27, 61
Leibniz	6, **145**, 158, 296, 380	Daniel Shapiro	368
Mary Leng	**275**, 284	Herbert Simon	**135**, 157
John Littlewood	**307**, 340	Brian Cantwell Smith	235
Lobachevsky	18	George Spencer Brown	**14**, 29, 114, 122, 125, 133, 160, **166**, **178**, 204-206, 341-352, **344**, 364-366, 368, 372, 380
Paul Lockhart	**165**, **174**, 180-181, **239**, 258, **272**, 283		
Danielle Macbeth	**11**, 28		
Ernst Mach	**207**, 235	Irving Stein	235
Alberto Martínez	**22**, **23**, 31, **54**, **57**, 64, **287**, 304	Ian Stewart	**271**, 283, **376**, 381
		Francisco Varela	**369**, 381
Henri Matisse	**14**, 29	F. Varela, E. Thompson & E. Rosch	**16**, 30
H. Maturana & F. Varela	**167**, 180		
Maxwell	18	Venn	10, 135, 217
Barry Mazur	**133**, 157	Viète	6
Joseph Mazur	**279**, 285	R. Vithal & O. Skovmose	28
Thomas McFarlane	**372**, 381	Hermann Weyl	**261**, 282
A. Michaelson & E. Morley	18	John Wheeler	**276**, 284
Randall Munroe (xkcd)	285	Alfred North Whitehead	**137**, 158
Napier	192, 296	A. Whitehead & B. Russell	**121**, **122**, **123**, 125, **197**, 205
Otto Neugebauer	**262**, 282		
Pascal	**xxiii**	Anthony Wilden	**164**, 179
Peano	4, 39, 58, 177, 223, 339, 371, 378	William Winn	368
		Ludwig Wittgenstein	206, **274**, 284
Charles Sanders Peirce	10, 28, **39**, 62, 135,	Stephan Wolfram	**22**, **26**, 31, 47, **87**, 116, 276, 284, 367

Index

SYMBOLIC CONCEPTS
primary definitions in bold
figure numbers follow their
page number in *bold italic*

ARITHMETIC

addition	**10**
by fiat	215
completed	63
base	262
base-2	263
base-10	262
base-e	173, 263, 300-301
explicit base	200, 263
implicit base	70, 249, 263, 265
uniform	262
exponents	53, 174, 338
power	251-255, 264, 267, 297, 338
power operator	167
rules of exponents	227-230, 346
fractions	291-297
group theory	59, 150, 202, 251, 339
associative	10, 59, 151, 159
commutative	59, 151, 159, 308
distributive	59
identity	59, 250
inverse	59, 151, 250
non-commutative	205
ring	344
history	5
abacus	5
body parts	5
counting table	5-6
doubling	84-85
knotted rope	5
Russian peasant multiplication	85
tally stick	5, 67, 84
without axioms	8
inverse	189
and/or	189
affirm/negate	189
double/half	189
even/odd	189
forall/thereexists	189
inverse object	250-251, 253
inverse function	189, 250-251
successor/predecessor	189
logarithm	299
antilogarithm	230
conversion	298-299
minus times minus	55, 57, 64, 205
non-Euclidean	6, 134, 280
Peano axioms	339
Presburger arithmetic	63
ratio	148, 295-297
tally	4, **33**, 83, 170, 203
unit	**34**

MATHEMATICS

abstraction	23-24, 203, 340
advanced	2
algebra	140
algorithm	6, 58, 81-82, 261, 268-269, 274-275, 277
ambiguity	159
arity	42, 126, 151, 153
axiom	8, **121**
axiom of choice	209
axiomatic style	122, 158-159
axiomatic system	8, 26
calculation	6, 277
Cartesian product	226, 342
calculus	**138**
numeric	26, 260, 276, 301, 375
pattern	133, 197, 302
predicate	338
propositional	13, 204
relational	342
category theory	11, 179
certainty	8, 306
chaos	24
closure	216-217
complete/consistent	217
complex number	263, 272, 280, 375, 380
computation	276-277
concrete	166
contextual expression	158
continuum	275-276
cultural imperialism	24
definition	123

Index

discipline	23
equal sign	123, 187
Euclid's *Elements*	157
fictionalism	**275**, 284
folly	203
formal theory	7, 17, 39
formalism	7, 21
geometry	7, 22, 124, 376
algebraic	158
Euclidean	6, 18, 135, **157**
fractal	53
non-Euclidean	6, 134
projective	193
grand unification	9
humor	29, 61, 280, 282, 285
induction	60, 234, 338
interpretation	121
knot theory	11
logic	26, 167
calculus	13, 204, 338
conjugate normal form	205
first-order logic	211
implication	259
map	121
matheism	24
manifold	179
metamathematics	378
native	21
nature abhors a vacuum	163
nominalism	16
ordered pair	295
partial ordering	141-**142**, 315
physics	19-20
Platonism	7, 16, 144, **158**, 206, 215, 237
purity	8
quantification	**187**
bound	187
domain	187, 216
existential	188
instantiated	188
universal	7, 187
quantum mechanics	19
relation	259-260
antisymmetric	142
irreflexive	14, 142
transitive	14, 142, 356
rigor	3, **17**, 135
self-evidence	120
set theory	8, 41, 167, 180, 209-211, 236
simplicity	22
space	151, 163, **165**
step	151
substitution	**49**
symbolic mathematics	2
symbolic formalization	7
theological	24, 27-28
theorem	**121**
truth	16-19
uncountable	271-277
variable	123, **141**, 186-189
variary	**42**
visual evidence	7, 9
wave equations	18

NUMBER

acrophonic	266
additive	134, 265
algebraic	266
Babylonia	261-262
common	5, 264
countable	210 **9-1**, 211-212
counting	207, 268
decimal	265
digits	262
Athenian	266
Arabic	21
Babylonian	261
Brahmi	8
Chinese rod	36
Egyptian	33
European	24
Mayan	39
Suzhou	23
factored	265
maximally factored	268
prime factors	263
formal operational definition	99
fraction	148-149
meaning	6
mixed	40, 263

393

Index

numerals	34, 261-262
one	166-168, 238
one-to-one correspondence	6, 34-39, 46, 62, **171**-172
order of magnitude	**67**
polynomial	265-268
real	271-277
algebraic reals	271
chaotic	274
Dedekind cut	**273**-275
irrational	263, 271
lawless	271, 274-275
transcendental	263, 271
Roman	67, 84
scientific	265
sensible	**279**-281
systems	261
cardinal	12, 264
complex	**263**
Conway numbers	154, 159
feasible	285
imaginary	263
integers	234
Kauffman numbers	67, 348-355
natural numbers	34, **161**, 170, 209-210, 233, 261-263, 276, 280, 288, 307, 363, 346
negative	263
one-two-many	277, 285
ordinal	63, 264
rational	263, 295
surreal numbers	154
whole numbers	24, 34-35, 38-**39**, 61, 71, 97, 170, 193, 201, 224, 258, 266, 284
uniform grouping	262
unit fraction	**263**
zero	151, **164**-165, 258-259, 262-263

REPRESENTATION

abbreviation	123
canonical	71-72, 205, 268
continuity	235
dimensionality	114-115
ellipsis	270, 315
icon	**3**, 139, 144-145, 212, 220, 238
iconic	1, 8
image	238, 240, 281, 305, 318
index	212
interpretation finger	25
language	**141**, 217
linear	268
meta-language	**162**
meta-symbol	123, 173
namespace management	335
naming	147, 283
notation	7, 9, 25, 37, 133, 200, 270, 303
place-value	262, 265, 269, 283
points	163
positional notation	262
Sapir-Whorf	340
semiotics	212, 219
sign	144, 212
structure	7
symbol	**3**
syntax/semantics barrier	7, 135, 369
syntactic sugar	34
textual	38-39
transcription map	**25**
typographical	36, 123-124, 134, 141, 146, 170, 262, 309, 343, 370

ICONIC CONCEPTS

BOUNDARY THINKING 1

beauty	1, 17-19, 30
belief	9
boundary form	11, 141
cognitive	9, 40, 114, 214 **9-3**, 237, **268**
category	13, 208, 213 **9-2**, 214-216, 218, 233 **9-6**, 236
construction	207-208, 210-**211**, 214, 218

Index

distinction	2, 5, 11-13, **14**-16, 218, 372			180-181, 210, 213-214, **218**, 235-237, 369, 377-378
load	3, 33, 66, 212	ensemble	12, 34, 170	
model	210 **9-1**, 213 **9-2**, 223, **231**, 233 **9-6**, 339, 370	environment	169, 325	
		existence	122, 126, 168, 171	
sensibility	**279**-280	existential graph	135	
shift	25-26, 84, 139	foundation	372	
computational load	66	goal/objective	1, 2, 5, 124, 155	
counting	83, 207, 211, **212**-223	holism	211, 236	
arithmetization	213 **9-2**, 231, 233 **9-6**	humane	**17**, 203, 371	
category	208, 214, 218	iconic	1-3, 6, 8, 17	
change-scale	223	calculi	**10**-11	
choreography	214 **9-3**, 215	concepts	34, 153-154	
count-by-pairs	222-223	math	26, 38	
domain	216	principles	122, 397 **15-2**	
ensemble	208	include the reader	13, 180, 309-310	
fusion	214	interaction	**211**-212, 218	
generalized	221-223	absence	138, 172	
identification	214	cerebral	144	
indication	214, 218-219	concrete	21-22	
remove reality	211-212, 233, 236	embodiment	11, 15, 308	
repetition	207	experiential	121, 324-329, 370	
replica	34, **207**	inhabitable	305	
replication	**219**-220	manipulative	109, 312, 329	
sequential	213-214, 220, 223	traversable	146, 305	
tally	203, 208, 218	spatial	136	
two-at-a-time	222	visceral	144, 151, 169	
what can be counted	210 **9-1**	knot theory	11, 364, 368	
design	26, 280, 314-315	linear	**115**, 211, 276, 310	
design choice	2, 369-370	structure	184, 209, 308	
diagrammatic math	10, 134-137, 339	textual	67, 71, 75, 134, 136	
diagrams	**136**	thinking	34, 135	
dialect	87, 145, 303	math education	115-116	
difference	14-**15**, 169, 377-378	advanced organization	119	
digital convergence	340	learn abstraction	21-23	
dimension	116, **144**, **307**, 359	loses information	85	
dimensional reduction	308	no math classes	22	
dimensionless	15, 163	novice learners	20, 367	
multidimensional	126, 144	PEMDAS	368	
one-dimensional	114, 307-308	professional tools	377	
two-dimensional	115, 307-308	symbol manipulation	58, 367	
three-dimensional	115, 307-308	totalitarian	9	
four-dimensional	307	unlearnable	83	
distinction	34, 121-122, 126, 139, 155, 167-170, 177,	why learn algorithms	66, 83-84	
		mereology	41	

395

Index

nesting	180-181, 268
object/process	126, 137, 148-**150**, 167, 211, 235-236, 250-251
objectivity	164, 310, 371
organic diversity	9, 28
postsymbolic	5, 27, 116, 306-307
sequential	126, 262, 270, 282
structure	42, 152, **159**
thinking	7, 134, 172, 202, 209, 266
simplicity	22-24
shares	61
unity	167
viewpoint	113-115
from above	113
from the side	113
orientation	309
point-of-view	310
systems	236
viewing perspective	13, 114, 150, 204-205, 317-318
you-are-here	321
virtual reality	27, 215
Wikipedia	154, 179, 365

CONTAINMENT 139

arity	126
boundary	16, 121, 248
brackets	123-124 *5-2*
angle	12, 61, 120, 124, **161**, 239-242, 244-247
capform	313, 327, 340
deconstruct brackets	311-314
double-struck round	124, 265
double-struck square	124
double-struck tortoise shell	47, 124
logic	124
parens	124
parenthesis	61, 63
set delimiter	62, 124
round	11, 84, 120, 124, **161**, 183
square	11, 61, 120, 124, **161**, 183, 191
tortoise-shell	36, 61, 85, 124, 170, 185
typographical	120, 161, 178, 343, 353
container	123, 126, 136
containment relation	1, 11, 123, 134, 137, 139, 150-**151**, 315
content/context	13, 126, 150, 158
crossing	126, 169
curvature	184
enclosure	312
exterior/interior	12, 16
fractured	157
generic	11
inside/outside	16, 39, 138, 166
interpretations of contains	66, **161**-162
ancestor	142
implies	139-140
is-a-member	139-140
parent-of	139-140, 142
rooted tree	139
shares-a-common-border	139
successor	139-140, 167
supported-by	139
Jordan curve theorem	29, 122
metaphor	16, 29
mutual independence	126, 137, 152
numbers	82
one relation	13-**14**, 36, 126, 139
open/close	157
outermost	34-**36**, 61-62, 85, 124, 141, **170**-171, 177, 185, 187, **310**, 320-323, 343-344, 349-350, 356
path	312
patterns of containment	137, 162
permeability	137, 246-248
J-transparency	131, 247, 302
opaque	246
pervasion	247
semipermeable	246, 259, 374
transparent	246-247
physical	12, 126
relation	312
schema	**16**
spaces	13
value-neutral	69, 85, 170
variables	123, 186-189

Index

ICONIC FORM

animated	145
animation	74-75, 82, 89-94, 97-99, 101-106, 108-109, 114
artifact	76, 268, 331
arity	42
cognitive	149, 233
dimension	114, 116, 308, 359
historical	118, 136, 209, 234
notation	85-86, **184**
perspective	113, 166, 204
axiom	126
cardinality	34, **172**, 208-209, 215
common boundaries cancel	71, 364, 368
idemposition	368
counting	83, 126, 207, 212-215, 268-269
equal	38, 62, **123**, 125-**126**, **168**, 187, 387
is-confused-with	123
not-equal sign	**123**, 125, 168, 170
permitted transformation	123, 126
substitute for	126
form	**121**, 126, 136, 175, 212, 225, 231
gradient	102, 322, 324
gravitational	99, 322
occlusion of shape	322
hybrid notation	**147**-148, 267
special symbols	147
models of multiplication	362 **14-9**, 364 **14-10**
arrangement	342, 361
grouping	362 **14-9**, 363
merging	362 **14-9**, 363
pattern-matching	361
substitution	342, 362 **14-9**, 363-364
touching	342
parallelism	44, 111, 152, 159, 202, 234, 262, 265, 270, 282
pattern	33
equation	183
rules	88
substitution	150
variable	186-187
put together	10, 126, 236
replica	**34**, 36, **207**, 238
structural transformation	5, 121, 126-131
structure sharing	101, 308, 319-320, 324, 332, 334, 340
unequal	126
unit	11-12
bounded nothing	166
empty container	12, 126, 137, 151, 166
units	34, 36
indistinguishable	34, 36, 39
no special	36, 165, 172
unit ensemble	26, 34-35, 59
whole numbers	35
varieties	**88**, **263**-265, **309**
geometric	309
relational	309
string	309
void	126, 151
absence	137, 151
blind to multiplicity	151-152, 175, 244
blind to sign	152
deletion	188
domain of non-existent forms	216
equality	**168**
illusion	186
meaningless	186
nature of things	163
non-concept	**15**
nothing	163-164
permeate	31, 137
two voids	380
typographical	136, 159, 314
void-based	**137**, 188, 378
void equality	47, 60, 63, **168**
void-equivalence	4, 126, **137**, 150-151, 186

MAPS **121**

accumulation	230 **9-4**
angle-brackets	256 **10-5**
arithmetic to logic	373 **15-1**
axiom systems	59 **2-8**

Index

binomial multiplication to
  network arrangement 105-106 *4-15*
factored number to
  depth-value   270 *11-3*
integers to
  block ensembles   109 *4-17*
  depth-value   77
  Kauffman string   353 *14-5*, 354 *14-6*
  unit ensembles   45 *2-4*
James spatial dialects 310 *13-1*, 311 *13-2*
James to
  bases   129, 143
  cardinality   290 *12-1*
  embedded bases   129, 143
  exponents   228
  fractions   294 *12-2*
  functions   120 *5-1*, 129, 139, *6-1*
  inverses   248 *10-1*
  logarithms   298 *12-3*
  multiplication   225
  numbers   129, 139 *6-1*, 240
  number types   263 *11-1*
  numeric encodings 265 *11-2*
  parallel arithmetic 129
  types of unit   161
James dialects to
  multiplication   146 *6-4*, 147 *4-5*
multiplication   362 *14-9*
numeric arithmetic to
  depth-value   88 *4-1*
  James   129, 143 *6-3*, 232 *9-5*, 278 *11-4*
  Kauffman string   341-342, 353 *14-5*, 354
  network ensembles 102 *4-10*
  Spencer Brown numbers 343, 345 *14-2*
place-value to depth-value 68 *3-1*
reflection to inverses 249 *10-2*, 252 *10-3*, 252 *10-4*
roadmap   310 *13-1*, 311, 320
spatial algebra   357 *14-7*
structural transformations 311 *13-2*
whole numbers to
  depth-value   72 *3-3*
  ensembles   35 *2-1*
  ensemble networks 99 *4-8*
  James units   171 *7-1*
  Kauffman numbers 67, 85, 348 *14-3*

Kauffman two boundary 350-351*14-4*, 352
Spencer Brown numbers 343 *14-1*

PRINCIPLES   121, 126, **378**
Accumulation   122, **171**
Additive   **3**, 33, 40-41, 60, 122, 134, 231, 238, 341, 348
Arrangement   105, **197**, 201, 224, 288
Axiomatic   **140**
Communality   **36**, 148, 208, 265, 288
Distinction   **15**, 121-122, 126, 167-169, 210, 352
Existence   122, 126, **168**, 177
General Semantics   213
Hume's   **62**, 122
Law of Calling   **372**-373
Law of Crossing   **372**-373
Multiplicative   **3**, 50, 122, 238
Participation   **13**, 106, 114, **135**, 149, 180, 283, 305, 340, 356, 378
Void   12, **15**, 126, 137, 151-152, 159, 162, 344
Void-Equivalence   126, **151**

ENSEMBLES

BLOCK ARITHMETIC 109-113
  addition   110 *4-18*
  digit facts   112
  division   113
  group and merge   109-111
  integers   109
  multiplication base-2 110-111 *4-19*
  multiplication base-10 111-112 *4-20*
  put   112
  standardize   109
  subtraction   110 *4-18*

CIRCLE ARITHMETIC   89-94
  addition   76 *3-6*, 89

398

Index

division 90 *4-2*
fluid boundary 89, 91-93, 116
fluid merge 91
multiplication 90 *4-2*, 92 *4-3*, 93 *4-4*
unit multiplication 91

DEPTH-VALUE 1, 65-69, 72, 87, 128, 262, 269

base 69, 262
cancel 88
canonical 71-75, 80, 88
  convenience 72
  depth-value **72**
  unit form 71
group 65, 69-**70**, 88, 128
  base-2 70
  base-10 70
  unit 65, 70
interpretation 88
  put 88
  substitute 88,
merge 69-**70**, 88, 128
order of magnitude 67
place-value comparison 68
reading 73
standardization 69, 72 *3-3*
uniform base 65, 266

ENSEMBLE ARITHMETIC 39-60

accumulation 37
addition 41 *2-2*
comparison 38, 58-59 *2-8*
ensemble **34**-35 *2-1*
idempotency 37, 346
fusion 41-43, 46, 59, 67, 69, 128, 209, 231
  fusion-bar 43
  lossy 43
  parallel 44
  void-based 43
multiplication 50-51 *2-6*, 52
partition 42 *2-3*, 43
polarity **45**
  angle-bracket **45**, 60
  anticommutativity 57
  multiplication 54 *2-7*, 55-57

negative unit 4 5, 54, 57
negative number 57
reflection 46, 59-60, 67, 128
  signed units 47, 54-55
  unit reflection 46, 77
spatial dialects 87
  dimension-free 88, 146
  one-dimension 88
  two-dimension 88
  three-dimension 88
substitution 47-53
  associativity **56**
  before-and-after 48
  commutativity **51**, 128
  construction **49**
  deletion **49**
  distribution 51
  division 52
  for 48-49, 51
  integrated 53
  into 48-49, 51-52, 54-55
  multiplication 67
  put 48-49, 51-52
  value maintenance 49
  unit reciprocal 53
tally arithmetic 33, 62
unit 34, 36
unit-ensemble 34-35, 60

ICONIC CALCULATOR 94-98

animation frames 89, 97-98 *4-7*
base mode 96 *4-6*
  base-1 ensemble 96
  base-2 binary 94, 96
  base-10 decimal 94, 96, 107
  base-10 digit 94-97
box dialect 94
cancel 95, 98
group and merge 94-95, 97
implementation 116-117
space mode 96 *4-6*, 97
  1D parens 96-97
  2D boxes 96-97
  3D blocks 96-97
ungroup 98
user interface 95 *4-5*

Index

NETWORK ARITHMETIC 99-109
  addition   101-102 *4-10*, 103 *4-11*
  arrangement   105 *4-14*
  base-2 numbers   99
  binomial multiplication   106
  cancel   101
  canonical   129, **148**
  cross-connect   102, 117-118
  decimal   107-108 *4-16*, 109
  dispersal   105
  factored form   103
  ground   **100**
  group and merge   100-101
  implementation   107
  multiplication   102-104 *4-12 4-13*, 105-106, 108 *4-16*, 117-118
  node   **100**
  reading   100
  self-similarity   101
  standardize   100
    parallel   101
  top   **100**
  variety   101, 106

PARENS ARITHMETIC 67, 74-81
  addition   74-76 *3-5*
  cancel   74-75
  canonical form   **74**
  create   74-75
  division   67, 74-81
  group   74-75
  group and merge   94-95, 97
  linear reordering   76
  merge   74-75, 85
  mixed polarity   **78**, 97
  multiplication   74, 80-81
  parallel   44, 94-95
    addition   76, 77 *3-7*, 79 *3-9*, 97
    group   74, 91, 97
    merge   71, 74, 76, 91, 97, 110, 113, 116
    put   74, 87
    substitute   91
    subtraction   79 *3-8*
  put   74, 81
  substitute   80-81
  subtraction   77-79
  ungroup   74-75
  unmerge   74-75

PATTERN EQUATIONS 121, 314-315
  accumulation   37
  arrangement   105 *4-14*
  fusion   43, 46, 59 *2-8*, 128
    void-based   43
    parallel   44
  group   70
    animation group/ungroup   75 *3-4*
    network   101 4-9
    unit   69 *3-2*, 70, 88 *4-1*, 128
  idempotency   37
  merge   69 *3-2*, 70, 88 *4-1*, 128
    animation merge/unmerge   75 *3-4*
    network   101 *4-9*
    parallel   71
  partition   43
  reflection   46, 77, 128
    animation cancel/create   75 *3-4*
    network   101 *4-9*
    unit   46, 77, 88 *4-1*
  substitution   47, **48**-53
    animation   75 *3-4*
    associativity   56, 59 *2-8*
    commutativity   51, 59 *2-8*, 128
    construct   49 *2-5*
    delete   49 *2-5*
    distribution over equality   49
    distribution over fusion   51, 59 *2-8*
    for-equality   49 *2-5*
    global   49 *2-5*
    identity   59 *2-8*
    into-equality   49 *2-5*
    inverse   59 *2-8*
    put-equality   49 *2-5*
    self   49 *2-5*
    value maintenance   49 *2-5*
  void-equality   47

Index

JAMES ALGEBRA

DIALECTS 145-147, 314-329
  blocks 146 *6-4*, 312, 325-326 *13-10*
  bounding box 145-146 *6-4*, 312, 315-316 *13-4*, 329-338
    calculation 331 *13-12*
    demonstration 336 *13-15*
    derivation 332 *13-13*
  bucket 312, 316-318, 317 *13-5*, 322
  map 146-147 *6-5*, 313, 320-322, 321 *13-7*
  network 146-147 *6-5*, 313, 318 *13-6*, 319-320, 329-338
    calculation 331 *13-12*
    demonstration 337 *13-16*
    derivation 333 *13-14*
  path 146-147 *6-5*, 314, 326-329, 327 *13-11*, 329-338
    calculation 331 *13-12*
    demonstration 337 *13-17*
    derivation 333 *13-14*
  example dialects 315 *13-3*
  stepping stone 322
  string 145-146 *6-4*
  room 146-147 *6-5*, 313, 324 *13-9*, 325
  types 311-314
    enclosure 312, 329
    path 312-314, 329
    relation 312-313, 329
  wall 313, 322-323 *13-8*

JAMES FORM 1, 121, 141
  accumulation 161, 170-172, 231-233, 287
  addition 126
  arrangement 1, 193-197, 224-226, 287
    factored 195-196
    generality 195
    generic 195
    multiplicative 195-196, 224, 226
    organizing principle 197

base 297-301
  arbitrary 148, 297
  base-e 173, 263, 300-301
  base-free 172-174
  embedded 248-250
  exponential 126, **243**, 269, 287
  fractional 300
  logarithmic 297-301
  calculation 232, 278, 329-332
  cardinality 171, 242, 287-290
  concepts 151
  demonstration 335-338
  derivation 332-335
  dominion 290
double boundary 185
double round boundary **173**
exponents 126, 183, 227-230
fraction 291-297
  addition 293-295
  compound fraction 294-**295**
  compound ratio 295-296
  multiplication 291-293
  ordered pair 295
  reduction 293
  ratio 296
  unit fraction 291-292
frames 121, 128, 196, **198**-199 *8-2*, 200-201
  arrangement 128, 198-199
  cardinality 128, **199**, 221, 234
  dominion 128, 242
  framed-content 198, 242
  frame-type 198
  generic **198**
  indication 199, **218**-219, 221, 234
  inversion 128
  J-conversion 128, 131, 302
  (J)-frame 271
  magnitude 128, **199**, 266-267
  unit magnitude 128, **200**, 269, 291
  void 128
infinity 177, 193, 197
  negative 183, 191, **202**, 239, 243, 257, 263
  non-numeric 4, 191, **243**
  positive 192, **263**

401

Index

James imaginary	263, 272
additive imaginary	4
J	**148**
J-transparency	302-303
logarithm of negative one	148, 156
generalized inverse	12, 156, 239
hash mark	**147**-148, **173**
hybrid	147
interpretation	142-143, 189-190
alternating	152, 190
power operator	167, 190, 228
logarithm	190, 192-193
inverse	250-257
add/subtract	253
angle-bracket	239
context	247-248
inverse object	250-251, 253
inverse function	189, 250-251
multiply/divide	253
power/logarithm	254
power/root	255
inversion pair	185, 189, 191 *8-1*, 253
multiplication	126, 224-226, 288
not repeated addition	225, 238
replication	224-225
logarithm	126, 183, 297-301
conversion	298-299
parallel	**129**
addition	286
arrangement	197, 293
counting	220-221
fraction	291, 302, 304
multiply	224, 288, 292-293, 302
promotion	288-289, 304
reflection	239-242, 301-303
imaginary	240
negative unit	240, 258
rotation	240
translation	240
repetition	170
replication	287-289
sign blind	303
simply nested	185
special symbols	147

unification	**175**-176
blind to multiplicity	175
domination	175
negative infinity	176
undefined for interpretation	176
unit	34, 151, 161
non-numeric	202, 259
variety	191 *8-1*

PATTERN EQUATIONS   121

arrangement	127, 140 *6-2*, 193, 198, 224
collect	193
disperse	193
parallel	197, 293
depth-value	128, 269, 291
dominion	127, 242, 244
equations	130
existence	127, 168
indication	127, 218
inversion	127, 140 *6-2*, 184, 251
clarify	184
enfold	184
void inversion	127, 162, 184
involution	127, 241
J-patterns	131
J-transparency	131, 247, 302
non-numeric	130
promotion	127, 244
reaction	127, 242
void-equivalent	258
reflection	127, 140 *6-2*, 241, 251
unit	127
void	240
replication	127, 219
separation	127, 241
two boundary	130
unification	175
unit accumulation	127, 170-171
void equality	168
void shell	170

Index

OTHER ICONIC SYSTEMS

BOUNDARY LOGIC   128, 133, 154, 159, 204-205, 259, 314, 340, 368, 372-375, 379-380

foundation	372
logical boundary	259, 372-374
pattern equations	373 *15-1*
arrangement	373 *15-1*, 374
calling	128, 372-373
crossing	128, 169, 352, 372-373
dominion	128, 373 *15-1*, 374
inversion	373 *15-1*, 374
involution	128, 373 *15-1*
pervasion	128, 259, 373 *15-1*, 375
unification	130, 175-176, 242, 352, 373 *15-1*

KAUFFMAN ARITHMETICS   67, 348-355

numbers in the form	348-350
context	349-350
fusion	348
multiplication	349-350
pattern equations	348 *14-3*
whole numbers	348 *14-3*
string arithmetic	342, 353-355
integers	353 *14-5*
interpretation	353 *14-5*
negative units	354 *14-6*, 355
notation	354-355
pattern equations	353 *14-5*, 354 *14-6*
subtraction	355
two boundary	350-352
additive content	350
core ideas	352
interpretation	350-351
multiplicative content	350
pattern equations	351 *14-4*
precursors of arithmetic	352

SPATIAL ALGEBRA   355-361, 367-368

distribution	358-360 *14-8*
factoring	358
labeled objects	357 *14-7*
partitions	357
physical blocks	356
space	356-357
empty	357
shared	357, 359
three-dimensional	357
touches	356-358

SPENCER BROWN NUMBERS   341-350

0-elements	343 *14-1*, 348, 350
1-elements	343 *14-1*, 347, 350
addition	343-344
exponentiation	346-347
interpretation	343, 345 *14-2*
multiplication	344
one relation	344
pattern equations	345 *14-2*
cancellation	345 *14-2*, 349-350
cross	343
null power	345 *14-2*
transfer axiom	345 *14-2*, 347
universe axiom	345 *14-2*
rules of exponents	346
Spencer Brown cross	342

VOLUME II   4, 48, 58, 62-63, 125, **130**, 140, 142, 156, 159, 168, 204, 209, 234, 258-259, 315, 378-379

abstraction	315
containers	204, 379
definition of number	58, 62, 276, 338
formalism	4, 339, 375
inverse	259
logic	159, 378-379
parallelism	234
pattern equations	125, **130**
equality	63, **130**, 168, 378
compose content	126, 130
compose context	126, 130
equality inversion	130
equality reflection	130
void equality	130, 168
two boundary	**130**, 156, 159, 257, 379
angle	257

Index

square	257
pattern-matching	48, 63
relations	140, 142

VOLUME III
4, 64, 125, **130**, **131**, 148, 156, 176, 193, 205, 233, 247, 257-258, 271, 295, 297, 303, 379-380

complex	**263**
exponents	233
i	4, 131
logarithms	297
π	131, 380
plane	258, 379
definition of J	131
divide-by-zero	156, 295, 378
exotic bases	193, 297, 379
J	4, 148, 156, 303, 379-380
J-frames	**131**
J-angle	131
J-conversion	13, 128, 131, 302
J-involution	131
J-self	131
(J) frame	271
non-numeric	**130**, 156, 161, 175, 242-244, 263
dominion II	130
infinite interpretation	130
infinitesimal	130
infinity	257, 378
indeterminacy axiom	30, 259, 290
square replication	130
square unit	130
unification axiom	130, 175, 242
pattern equations	125, **131**
J-parity	**131**, 380
J-parity whole	131
J-parity part	131
J-theorems	**131**
J-absorption	131
J-conversion	128, 131, 302
J-occlusion	131
J-reflection	131
J-self-inverse	131
J-self-occlusion	131
J-transparency	131, 247, 302
J-void object	131
J-void process	131
J-void rally	131
multiplication of signs	63, 205

SYMBOLS AND ICONS

1,2,3...	whole number domain	39, 61, 123, **177**, 212, 222-223, 276
●	dot unit	34
◇	inverse unit	**45**, 54, 77
o	James unit	124, 139, **167**
●●●●	four tally	**39**, 72
oooo	four tally	141, 147
/////	five tally	26, 33
...	ellipsis	**177**, 197, 219, 270
..N..	cardinality N	**177**
ℝ	real domain	**276**
∞	infinity	161, **176**, 239, 257
( )	round	139, **161**
[ ]	square	139, **161**, 259
< >	angle	139, **161**
( )	value-neutral shell	**36**, 177
⟨ ⟩	logic boundary	124, 128, 259, **373**-375
(( ))	unit magnitude	124, 128, 200, 265, **269**, **291**
[ ]	substitution	**47**
[ ]	two-boundary	130, **257**-258
{ }	set delimiter	62, **124**
(( ))	James base	**173**, 249, 291, 299
([ ])	inversion pair	184
[( )]	inversion pair	184
[<( )>]	J	129, **148**, 156
=	equal sign	**123**, **168**, 187
≠	difference sign	**168**
☞	interpretation finger	**25**
⇒	process arrow	**43**
⇔	equation equality	**123**
#	hash-mark	147-148, **173**, 177
void	non-symbol	**162**, 177
⌐	Spencer Brown cross	**342**

404

Index

COVER

page 332
*path dialect
reflecting on its own
network dialect form*

page 310
map dialect

page 93
*fluid circle
dialect*

page 109
*standardize
base-2 block
numbers*

page 41
*ensemble
fusion*

page 13
*form as
environment*

page 324
*room dialect
Arrangement axiom*

page 122
Jordan curve

page 11
*boundaries
both separate
and connect*

COVER WORDS

Arithmetic evolves.

The Laws of Algebra are *design choices*
made before the Internet, before computers, before electricity.

The arbitrary symbols we memorize and try to recall
are abstract, one-dimensional, sequential, detached.

Iconic arithmetic is physical, sensual, natural.

Tangible mathematics connects us
to our cultural and biological heritage,
to the authentic world of experience.

Icons look and feel like what they mean:
embodied, multidimensional, concurrent, inclusive.

Arithmetic has evolved.

Index

WEBSITES

iconicmath.com	60, 82, 89, 159
iconicmath.com/arithmetic/blocks/	109
iconicmath.com/arithmetic/containers/	82, 89
iconicmath.com/arithmetic/depthvalue/	82
iconicmath.com/arithmetic/networks/	99, 114
iconicmath.com/arithmetic/parens/	82
iconicmath.com/arithmetic/units/	60
iconicmath.com/algebra/containers/	60
iconicmath.com/algebra/spatial/	361
iconicmath.com/calculators/binary/	94
iconicmath.com/calculators/decimalunits/	94
iconicmath.com/calculators/digit/	94
iconicmath.com/calculators/iconiccalculator/	82, 94
iconicmath.com/logic/boundary/	380
arxiv.org/pdf/math/0112266.pdf	368
cs.nyu.edu/davise/personal/PJDBib.html	28
homepages.math.uic.edu/~kauffman/Laws.pdf	366
homepages.math.uic.edu/~kauffman/NUM.html	62
homepages.math.uic.edu/~kauffman/Papers.htm	368
homepages.warwick.ac.uk/staff/David.Tall/pdfs/dot1991h-gray-procept-pme.pdf	158
pauli.uni-muenster.de/~munsteg/arnold.html	159
socialsciences.exeter.ac.uk/education/research/centres/stem/publications/pmej/pome19/index.htm	62
www.computerhistory.org/revolution/calculators/1	27
www.csc.liv.ac.uk/~sazonov	285
www.dpmms.cam.ac.uk/~wtg10/philosophy.html	118
www.math.fsu.edu/~wxm/Arnold.htm	159
www.integralscience.org/tom/	381
www.stephenwolfram.com/publications/mathematical-notation-past-future/	117
www.webofstories.com/play/michael.atiyah/89	235
xkcd.com/899/	285

Printed in Poland
by Amazon Fulfillment
Poland Sp. z o.o., Wrocław